The New Engineering Game

Strategies for Smart Product Engineering

Tim Weilkiens

The New Engineering Game

Strategies for Smart Product Engineering

Tim Weilkiens

ISBN 978-3-9818529-4-3

Publisher: Tim Weilkiens

Contents

The New Engineering Game - Strategies for Smart Product Engineering by Tim Weilkiens

/MBSE4U

www.mbse4u.com

Published by MBSE4U

First Edition, First Printing 2018

Print edition printed by Lulu.

Print edition: ISBN 978-3-9818529-4-3
PDF edition: ISBN 978-3-9818529-5-0
ePub edition: ISBN 978-3-9818529-6-7
MOBI edition: ISBN 978-3-9818529-7-4

About MBSE4U

Publishing on the Pulse of the Market

MBSE4U - Tim Weilkiens is my publishing organization for MBSE books that are regularly updated to follow the dynamic changes in the MBSE community and the markets.

MBSE4U has published

- Benjamin Weinert. Ein Framework zur Architekturbeschreibung von sozio-technischen maritimen Systemen. 2018.
- Christian Neureiter. A Domain-Specific, Model Driven Engineering Approach for Systems Engineering in the Smart Grid. 2017.
- Tim Weilkiens. Variant Modeling with SysML. 2016.
- Tim Weilkiens. SYSMOD - The Systems Modeling Toolbox - Pragmatic MBSE with SysML. 2nd edition. 2016.

Please let me know if you want to write a book published by MBSE4U (tim@mbse4.com).

⁄MBSE4U

About Tim Weilkiens

Tim Weilkiens is a member of the executive board of the German consulting company oose, a consultant and trainer, and active member of the OMG and INCOSE community.

He has written sections of the initial SysML specification and is still active in the ongoing work on SysML. Tim is involved in many MBSE activities, and you can meet him at several conferences about MBSE and related topics.

As a consultant, he has advised many companies in different domains. His insights into their challenges are one source of his experience that he shares in books and presentations.

Tim has written many books about modeling including *Systems Engineering with SysML* (Morgan Kaufmann, 2008) and *Model-Based System Architecture* (Wiley, 2015). He is the editor of the pragmatic and independent MBSE methodology SYSMOD – the Systems Modeling Toolbox [We15].

You can contact him at tim@mbse4u.com and read his blog about MBSE at www.model-based-systems-engineering.com.

History and Outlook

This chapter gives a brief overview of the version history of the book and looks forward to future edits.

Version 1.0 - Initial version

Targeted topics for future versions:

- Impact of digitalization on the society
- Additional characteristics of the new engineering game
- Updated and additional methods & tools

Preface

As a consultant for model-based engineering, I have accompanied many organizations from different domains on their way to meet the challenges of complex and dynamic markets. There are many publications around that covers the challenges from a high-level perspective. This book is intended to close the gap between the high-level visions and the daily engineering tasks. It provides strategies to align the engineering of products in a purposeful and fruitful direction.

A disadvantage of classical books is the low frequency of updates. That was my motivation to publish this book. I continuously update the toolbox based on feedbacks and experiences from industrial projects and changes in the engineering community.

A disadvantage of self-published books is the missing quality gate of a traditional publisher. There is no copy-editor that for instance proves the correct usage of the English language - in particular, if the author is not a native speaker - or if the line of arguments always makes sense for the readers.

I appreciate any feedback on the book. Be it on the content or my English skills. You can reach me by email: tim@mbse4u.com.

I like to write a book in a gender-fair language. On the other hand, I avoid cluttering the flow of reading by always using both genders in the same sentence. Therefore I have only used one gender where it was not appropriate to use gender-neutral language. Feel free to replace the gender with your favorite one wherever it is appropriate.

I want to thank you for buying this book. You spent your money well. Now you have a toolbox for smart product engineering. Moreover, I have some money to finance the infrastructure to provide more engineering information to the community.

If you need training or consulting services feel free to contact me. My company - the consultancy oose - provides training and coaching services for systems and software engineering.

Tim Weilkiens, December 2018, tim@mbse4u.com

1. Introduction

The engineering of products faces an increasingly complex and dynamic environment. Megatrends like the Internet of Things (IoT) and the industrial internet respectively *Industry 4.0* change the rules on the field. They require a new kind of engineering and thinking. Only companies that can adapt themselves to the frequent and partially disruptive changes in the complex and dynamic markets are successful in the long term.

First signs of this new playfield can be observed, for example, from the rise and steep fall of Motorola and Nokia or the tremendous success of Google and Apple. Tremendous things happen in very short times. The markets are highly dynamic.

Nearly every day you can find a remarkable hint of the change in the news. If you thought Paypal was a revolution in the banking sector, you should have a look on SnapCash [Sn16]. SnapCash transfers money the moment you enter a sum of money in a SnapChat dialog. Traditional money transfers can last days even if the same bank manages the accounts. Moreover, this was remarkable when I wrote this book. It is probably already an old hat when you read these lines.

Speaking of Paypal, the founder Elon Musk is another example of disruptive changes. Traditional automotive vendors greeted his idea of building electric cars with smiles. Now his Tesla sells better than traditional luxury cars in Western Europe [LAT16]. Same for SpaceX.

There are many publications available that capture the high-level viewpoints of the impact of the Internet of Things, Industry 4.0 and Co. This book closes the gap between the high-level reflections and the need for daily engineering methods and tools to face the challenges.

After an introduction and motivation of the theme, the book describes the first three industrial revolutions and their consequences and concludes with the predicted fourth industrial revolution. In the light of the fourth industrial revolution, the second chapter depicts the need for a new kind of engineering. The third chapter provides valuable principles, patterns, methods, and tools for engineering organizations to be successful on the new engineering playfield.

This book covers many exciting topics like business model canvas, model-based systems engineering, industry 4.0, industrial internet, and much more. Each topic is worth a book of its own and those books are already available.

Thus I only give an overview of the topics to understand the concepts and to be able to do the first steps. Besides the worth knowing overviews, the book depicts the relationships between the concepts.

This book fills up your toolbox to master the new engineering game.

2. The Industrial Revolutions

A revolution is a dramatic and wide-reaching change in conditions, attitudes, or operation [Ox17]. People and organizations who do not align with the change have a good chance not to survive. Thinking of revolutions most people have a forcible overthrow in mind that happens many years ago. For example, the storming of the Bastille during the French Revolution on 14 July 1789 (figure 2.1).

Figure 2.1: Storming of the Bastille on 14 July 1789 [ImgSB]

Nowadays, we still have revolutions. For example, in the early 2010s the Arab Spring affecting Tunisia, Egypt, Libya, Yemen, Syria, and Bahrain [Br15].

Changing the focus from politics and overthrows of governments to the industry and engineering domain, we also find
revolutions with dramatic and wide-reaching changes.

Industry 4.0 describes the radical change in the manufacturing
domain. It originates from the German project *Industrie 4.0*
that is part of the *High Tech Strategy* of the German government [Ger14]. The German term *Industrie 4.0* points to the
fourth industrial revolution.

That implies that we already had three industrial revolutions
before. The following sections present and discuss the first
three industrial revolutions that finally result in the fourth
industrial revolution.

Figure 2.2 gives an overview of the industrial revolutions. It
is based on a similar figure in the paper *Securing the future
of German manufacturing industry: Recommendations for
implementing the strategic initiative INDUSTRIE 4.0 - Final
report of the Industrie 4.0 Working Group* [I413].

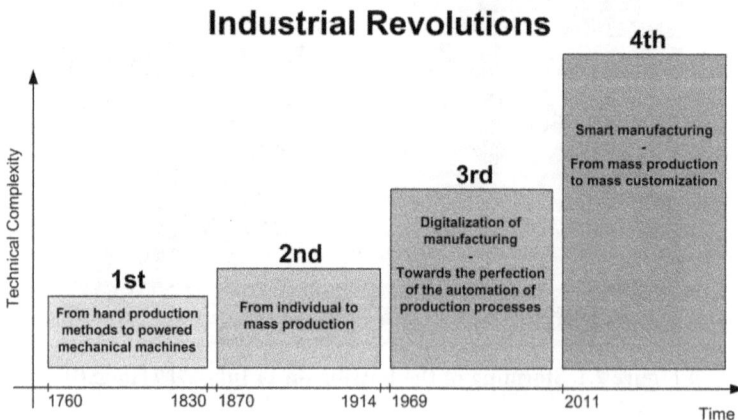

Figure 2.2: The Four Industrial Revolutions

Sometimes people talk about only three industrial revolutions. In their context, the third industrial revolution is what here is called the fourth industrial revolution. The four revolutions are the German view of the industrial landscape. In the USA was no industrial revolution what the Germans call the third industrial revolution. Therefore, the Americans count in summary only three revolutions (e.g., Rif13). You find details about the different viewpoints in section 2.3 about the third industrial revolutions.

I call the third and the fourth industrial revolutions also the quiet revolutions. They do not come along with prominent events, or people like other revolutions did. The change happened respectively still happens quietly step by step. Nevertheless, the result is a disruptive change of the industrial landscape.

The historical review shows the effects of industrial revolutions and general development. Against this background, it is easier to look at the current development from an out of the box perspective. The retrospect shows that the technical progress is inevitable. You cannot stop progress. Moreover, it is waste to spend effort on trying to stop it. Instead, you should spend your valuable time and resources to guide the progress in the right direction and to adopt the changes that accompany the progress.

However, the technical progress is only one side of the coin. Just as important or even more important is the role of the people.

2.1 The First Industrial Revolution

The first industrial revolution took place in the 18th century from 1760 to 1830 [As98]. The production of things changed from hand production to mechanical production machines in combination with the usage of power.

An important role played the power loom (weaving machine) invented by Edmund Cartwright 1784 (figure 2.3). Steam powered the weaving machine. Although it improved the weaving process, it took some time before the technology was widely spread.

The weaving machines raised many concerns about losing workplaces and worsened the conditions for the workers. Finally, it led to the so-called *Machine Storm*. The protest movement destroyed many production sites. Nevertheless, the power loom technology was being improved over the next 47 years until a design by Kenworthy and Bullough made the operation completely automatic.

The change from hand production, not only in the weaving domain, to powered factories changed the conditions for the workers dramatically. Increased unemployment, low wages, long working hours in hot and dusty factories. That was the beginning of unions and similar organizations to fight for more rights and better conditions for the workers.

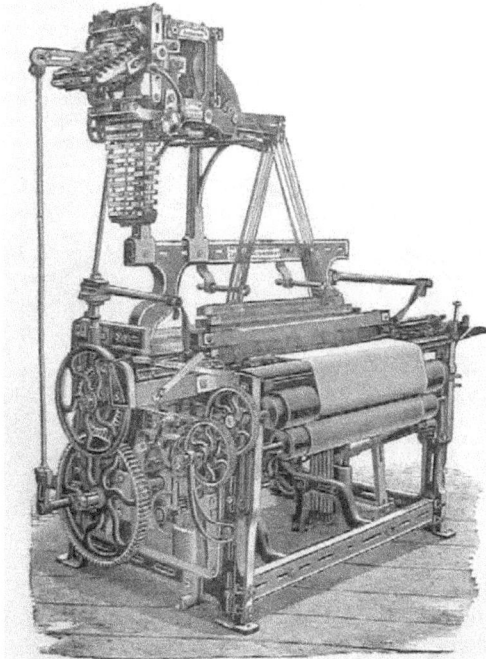

Figure 2.3: Power loom [ImgPL]

The construction of factories also centralized the previously widely distributed working places and caused an increased centralization and urbanization.

The agriculture was another area affected by the change. Horse-powered threshing machines (figure 2.4) replaced workers and has also lead to a protest movement. The machines were only one piece of the whole story. The wages and other social issues have been damned up and heated up the riots.

A famous protest movement was the *Swing Riots*. The name originates from the fictitious *Captain Swing*. He was the head of the Swing Riots, and the signee of the letters sent to farmers, government people and others who were seen responsible for the situation. The Britain government cracked down on the riots and transported more than 500 people to Australia, 600 people imprisoned and 19 people executed [Na16].

Figure 2.4: Horse-powered threshing machine [ImgTM]

However, the progress of using machines in manufacturing was unstoppable: individual craftsmanship distributed over the countries turned into centralized factories with powered production machines. The change of the production landscape created much new employment and destroyed much other employment at the same time.

The factories were much more efficient than the distributed local workshops. The price for the efficiency was less flexible production processes. That was not a big problem during that time. Since the markets were thirsty for products, there was not much demand for considering individual needs. The people were happy about the things they got from the factories.

That was the first step towards the era of mass production that dominated the second industrial revolution.

2.2 The Second Industrial Revolution

The second industrial revolution took place from 1870 to the beginning of the first world war in 1914 [En16].

Violence and destruction did not accompany the second industrial revolution like the first one. It had many aspects. Steam-powered railways and ships enabled a broad distribution of goods, people and ideas. The invention also drove the latter and the spreading of telegraphy and the telephone technology. It was a first wave of the globalization phenomenon explored in more detail in chapter 3.1.

The most prominent aspect was the introduction of the principle of assembly line production that again changed the landscape of the industry. The assembly line production enables the mass production and massive reduction of production costs.

Henry Ford (figure 2.6) and his assembly line for the production of the car *Model T* (figure 2.7) is typically mentioned as an example for the beginning of the mass production era.

It should be noted that Henry Ford was not the inventor of mass production, although you quickly get the impression when reading publications about his era. He combined existing concepts and was powerful enough to influence the industry to be remembered in history.

Figure 2.6: Henry Ford (1863-1947) [ImgHF]

Mass production can be observed many times in history. For example, the production of crossbows in China during the Warring States period (475 - 221 BC) [Pe96]. The assembly lines of the slaughterhouses in Cinncinatti are another example and one of the first spots of the second industrial revolution in 1870 [I413].

Figure 2.7: Ford assembly line [ImgFA]

Henry Ford was a practitioner and industrialist. Frederick Winslow Taylor (figure 2.8) provided the theoretical background of his approach. For some time Taylor worked for Ford to improve his production processes. Taylor was a mechanical engineer. He wanted to improve the efficiency of the production process and is regarded as the father of *Scientific Management*, also called *Taylorism*.

Figure 2.8: Frederick Winslow Taylor (1856-1915) [ImgFT]

Scientific Management is a management concept based on a scientific approach. One component of scientific management is the measurement of working times to determine efficiency leaks, to reorganize processes, and to define schedules for the individual working steps. Taylor observed the production of Ford to determine more effective production steps.

Another component of scientific management is the transfer of control from the workers to the management. The managers plan the work and specify how the work is to be performed. The workers only execute the tasks. It should be mentioned that Taylor had not a good opinion of workers. He was convinced that the workers have no understanding of what they were doing [Mo89]. However, for the sake of efficiency, he also proposed good things for the workers, for example, regular rest breaks to relax. He observed that regular breaks increased the quantity and quality of the output [Tay11].

The separation of control and work lasts until today. Interestingly, it has started to turn back and to merge control and work together again. I have looked at this aspect in chapter 3.7.

Taylor would be happy to know that his work paved the way for the automation of the third industrial revolution and the outsourcing in the era of globalization. The separation of work processes into discrete units in scientific management sharpened the workpieces for automation and outsourcing.

The increased worse conditions for the workers led to protest and strikes. Ford was aware of this and introduced the 5 Dollar and 8 hours working day. During his time typical wages were 2,5 Dollar and 12 working hours per day. At the same time, he could decrease the price of his car Model T from 870 Dollar to 270 Dollar [He16].

The producers of goods drove the markets. As long as the markets were not mature, the people were pleased to get new goods for affordable prices. They bought what was provided by the producers. That changed in the more recent industrial revolutions from producer-dominated markets to customer-dominated markets.

The second industrial revolution ended with the beginning of the first world war in 1914. The first world war was also a dramatic and wide-reaching change, but of an entirely different category.

We can observe the same patterns as during the first industrial revolution. A trigger and enabler for the revolution was a change in energy and communication technologies. For the first industrial revolution, it was mainly the availability of the steam-based energy for machines. For the second industrial revolution, it was electrical energy and in particular a dramatic improvement of communication channels like the telegraph.

Again, the technical progress was unstoppable and herewith the change of the industrial landscape. Because of the latter, we call it a revolution. The victims of the revolutions were the industrial workers.

Now and then it is a challenge for the society and the governments to find a good path through an industrial revolution for all layers of the population. Stopping the progress is not an option.

2.3 The Third Industrial Revolution

The third industrial revolution was also not accompanied by protests, riots, and violence. It happens step by step without a clear start and end point.

Some sources call the third industrial revolution what is called the fourth industrial revolution in the context of *Industry 4.0* (e.g., Ec16 and Rif13). The Germans have optimized their production, and that leads to the third industrial revolution in Germany. In the USA the production was more outsourced than optimized during that time. Therefore, they had no industrial revolution in the traditional manufacturing industry. However, they had another kind of a revolution: the rise of the software industry. The result was the outstanding success of companies like IBM or Microsoft and Google or Apple. However, that revolution has gotten no name, and they skipped so to say one industrial revolution. Therefore we have a slight confusion with the numbering of the industrial revolutions.

The fourth industrial revolution - described in detail in the next section 2.4 - is about the interconnection of production machines, resources, and consumers. The third industrial revolution is about the digitalization of the manufacturing.

Nowadays, people often use the term *digitalization* for a different meaning. Namely, the transformation process that happens during the fourth industrial revolution. Initially, it merely means the transformation of analogous signals to a digital format for the processing by a computer.

Digitalization and connectivity go hand in hand. If technology enables more connectivity, it leads to more digitalization. For example, the digitalization of things in the world of the Internet of Things (IoT). In summary, it can be said that the term digitalization is widely used to denote the changes in the fourth industrial revolution. At the same time, its original meaning fits more with the third industrial revolution.

The third industrial revolution started in the 1970s with the introduction of programmable logic controllers (PLC) to automate production processes [I413]. Dick Morely is regarded as the father of PLCs. He and Georg Schwenk founded the company Bedford Associates in New England, USA, in 1964. Bedford was a control system engineering company. Their product *Modicon 084* was the worldwide first PLC.

Again, it was not a single stroke of genius. Instead, the time was ripe for the invention of PLCs. Control systems already managed production machines. A control system controls conditions by monitoring them, calculate necessary adaptions and change the conditions with actuators if necessary, and repeats that cycle again and again (figure 2.8). Morely wanted to overcome the long and intensive programming and debugging of control systems before they went into operation.

At the same time General Motors Hydramatic division - a producer of transmissions - struggled with their control systems at the production line. In consequence, they wrote a specification of a standard machine controller that should solve all the problems. Seven companies got the specification, and three of them delivered prototypes. Amongst them Bedford Associates with their Modicon 084 PLC system. All three prototypes were installed and intensively tested in operation. Later the Modicon 084 became the preferred PLC for factory engineers.

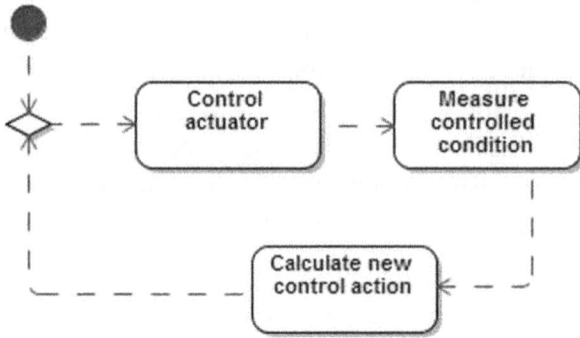

Figure 2.8: Principle of a Control System

A more detailed description of the history of PLC can be found in the article *The Dawn of the Programmable Logic Controller (PLC)* by Ken Ball published by the PULSE magazine [Ba15].

The PLCs were getting more and more powerful, and the automation level of the manufacturing processes increased steadily. Although the triumph of the PLCs began in the US, it was primarily the Europeans and in particular, the Germans who perfected the automation of manufacturing. The US industry had massively outsourced traditional manufacturing factories. The outcome of the third industrial revolution was an extremely optimization of the manufacturing processes according to quantity, quality, and costs.

The transition from the third industrial revolution to a post-phase or the beginning of the fourth industrial revolution was a fluid transition.

2.4 The Fourth Industrial Revolution

As for the third industrial revolution, it is hard to say when the fourth industrial revolution has started or even if it has started already.

The unique thing about the fourth industrial revolution is that it is a predicted revolution. The first, second, and third industrial revolutions were afterward observed as a revolution. For the first time, we have the chance to act in the sense of the change. It is an excellent opportunity for companies.

The era of mass production and extremely optimized processes have led to effective, but inflexible organizations. That is no problem as long as those organizations operate in a stable environment. Nowadays, we have highly dynamic markets, and the environments of many companies are anything but stable. The markets of the fourth industrial revolution require flexibility and adaptability.

The primary technical enablers of the fourth industrial revolution are the cyber-physical systems (CPS) (section 3.4). After the Internet of People (IoP), they bring things into the virtual world. Now everything is connected, and the real world merges with the virtual world. The IoP merges with the Internet of Things (IoT) to the Internet of Everything (IoE).

In manufacturing, the production machines create an intelligent network along the value chain. They communicate with the products itself, with logistic systems, or with other production machines. On that basis, complex processes can be automated and adapted in real time to current conditions and needs. Production machines can order their supplies, product pieces can individually tell the production machines how they should work on them, and so on. There is immense potential for new and optimized manufacturing processes.

The connection of things in combination with significant progress in data science and machine learning enables machines to do more and more brainwork. They can make decisions on a case-by-case basis and be a driving part of complex processes. That makes production processes more intelligent and enables the shift from mass production to mass customization. You still have high batch sizes, but every single product can be customized to specific needs.

Figure 2.9 depicts the return of complexity.

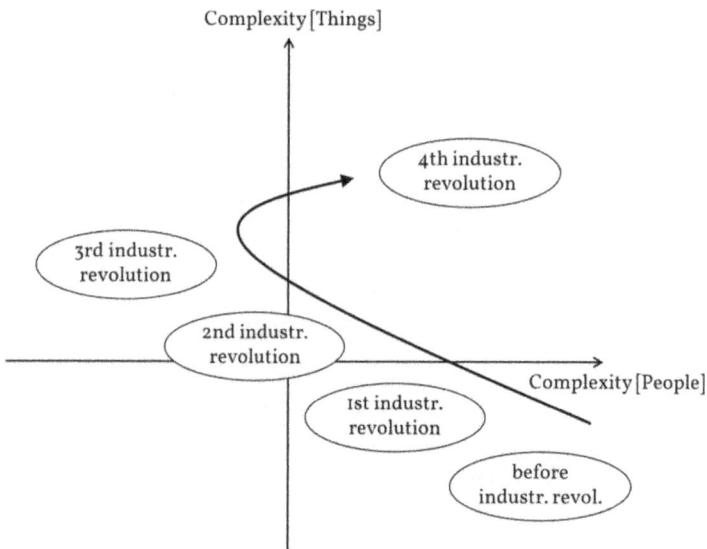

Figure 2.9: Return of Complexity to the People

The first three industrial revolutions decrease the required craftsmanship of people by automating more and more tasks. At the same time the complexity of things - the products and the manufacturing machines - increased. With the fourth industrial revolution, we have reached a turning point. On one side the machines can undertake more fuzzy tasks and make more "intelligent" decisions. On the other side, the development and orchestration of the appropriate engineering and manufacturing environment require more craftsmanship again.

The term Industry 4.0 has its origin in an initiative of the German government. As a world's leading manufacturing equipment supplier it is of high interest to the German government to strengthen the manufacturing industry and to prepare the path for ongoing future success. The prediction of the fourth industrial revolution and the term *Industry 4.0* was first used in 2011 at the Hannover fair in Germany. *Industry 4.0* is an essential project for Germany to be an outstanding manufacturing location as a high-wage country in the global market. Besides the specific interest of the German government, it is, of course, important for all companies in the manufacturing domain worldwide. Other countries have started similar initiatives, for example, *Made in China 2025*.

Industry 4.0 is also called *Smart Manufacturing* in other regions of the world. The Smart Manufacturing Leadership Coalition (SMLC) defines smart manufacturing as *"the ability to solve existing and future problems via an open infrastructure that allows solutions to be implemented at the speed of business while creating advantaged value."* [SMLC17].

The National Institute for Standards and Technology (NIST) has a more technical viewpoint and defines smart manufacturing as "systems that are fully-integrated, collaborative manufacturing systems that respond in real time to meet changing demands and conditions in the factory, in the supply network, and in customer needs." [SMLC17].

Many terms are floating around for Industry 4.0 or the technologies within the context of Industry 4.0:

- Smart Manufacturing
- Internet of Things (IoT)
- Cyber-physical systems (CPS)
- Big Data, Data Science
- Machine Learning
- ...

The German Industry 4.0 initiative has a strong focus on manufacturing. It is about the connection between production machines and the products with each other and the usage of a massive amount of data produced by sensors placed all around the production process. However, this is only one part of the change. Nearly every domain is affected by it.

The IoP has very lively shown how technology changes the world. With a small device called a smartphone, we are now able to communicate anywhere with people all over the world. Moreover, it is not only a one-to-one communication like a telephone. It is a many-to-many communication and connects us with other communities and cultures. The Arab Spring revolution was an excellent example of the power of the internet of people. It was also a revolution - a dramatic and wide-reaching change. Now technology enables to connect nearly everything - people and things - with the internet.

Products can be connected with their vendors and report on their status. That data can be used to improve the next versions. Another scenario is predictive maintenance where the data - collected by the products - is used to predict the next maintenance work. That can reduce maintenance costs because it is only done when necessary. Moreover, it enables complete new business models, for example, Rolls-Royce sells thrust instead of aircraft turbines [Ec17]. In particular, that business model gained notoriety with flight MH370. After the flight disappeared from the radar screens, the turbines still sent data to the vendor and witnessed that the flight was in the air for four more hours [Ind17].

The fourth industrial revolution is not only about manufacturing. It is an overall change in all product life cycles and business models.

3. The Context of the New Engineering Game

The markets are changing dramatically. Some engineers think that it has no significant impact on their work. Market stuff is for managers and business people. They are wrong!

The classical engineering world does not work well with the new challenges of the complex and dynamic markets. Job descriptions are changing, new kinds of jobs were created, and complete new development disciplines come up. All along the markets request more and more innovative products. Innovation and complexity are tightly coupled.

This chapter reveals the changes that have significant impacts on the engineering disciplines. First, we touch the globalization in section 3.1. An unstoppable megatrend that changed and still changes nearly any aspect of our lives.

Mega challenges are the dynamics and complexity of the markets, products, and engineering organizations. The next section 3.2 explains the meaning of *dynamic* and *complex* in the context of this book.

Organizations and products are changing rapidly and extensively. That makes Conway's Law very interesting. It describes the dependency between products and the organizations that develop them. Section 3.3 presents Conway's Law and discusses the relevance for engineering organizations.

A new kind of systems are the *Cyber-physical systems* (CPS). They are often mentioned in the context of the Internet of Things (IoT) and are taken as an example of disruptive product innovation. Section 3.4 covers these new figures on the engineering playfield.

The borders of functional responsibility between the specialty engineering units like software, mechanical, electrical, and so on are getting more and more blurry. Section 3.5 covers the new kind of interdisciplinary engineering.

Typically, we have a strong focus on processes in engineering. However, the best processes are not successful without experienced people who perform them. Section 3.6 deals with the resurrection of craftsmanship.

Also, the way we work together is changing. It is a sideshow for the engineering game, but significant enough to briefly cover the New Work World movement in section 3.7.

Change costs effort, and you would not do it until you see the real value or you feel a pain that kicks you into the change. Section 3.8 calls that the *Gap of Slackness* and discusses how to bridge the gap.

A change does not only cost time and money. It also needs much power from the people who are inside the changing system. Moreover, sometimes it costs too much power. Finally, we do a brief excursion about project heroes and burnouts in section 3.9.

The following chapter 4 lists some valuable moves for the new engineering game.

3.1 Globalization

The globalization is a megatrend since the past 15 years. The term designates the global communication of people, exchange of products, and different views like culture, politics, general ideas, and more. Enablers of the globalization are transportation and communication networks. In the past, some times also got the label *globalization*, for example, the second industrial revolution (chapter 2.2). The internet has boosted the globalization since the 1990ies. That is the era that changes the rules and the playfield of the engineering disciplines.

There is no single universal definition of the term globalization. Some people say that it is not possible to give a definition. The paper *Definitions of Globalization: A Comprehensive Overview and a Proposed Definition* by Dr. Al-Rhodan and Gérard Stoudmann [Al06] gives a good overview of different attempts to define the term. Based on their analysis they derived their definition of globalization:

Globalization is a process that encompasses the causes, course, and consequences of transnational and transcultural integration of human and non-human activities. [Al06]

In any case, you can observe that the increased communication and international collaboration has a dramatic impact. Today, we have small devices - the smartphones - that connect us with other people all over the world. It is not just a single point to point connection like the standard phone. It is a gateway to connect cultures, to witness events, and to exchange ideas. The smartphone enables incredible opportunities and dynamics. For sure, it was one crucial part of the Arabic spring revolution (see also section 2.4).

On the first view, it seems that the globalization enlarges the market and increases sales. On the second view, you recognize that it is the reverse direction. The globalization narrows the markets. Independently of where your business is located, you can produce and sell your products worldwide. Your competitors can be everywhere. Different countries and cultures have different cost structures, capabilities, and ideas. In many domains, the customers do not care where their products were produced and where the producing company is located. Important are the price, the quality, and the time-to-market.

The transportation logistics and communication channels are well established. The same way you can produce and sell products in any region of the world, other companies - your competitors - can do the same. That leads to enormous pressure on the companies. There is always a risk that another company delivers faster, cheaper, or with better quality, or all together at once. The global markets guarantee if you were not able to satisfy the markets needs someone else does it.

The companies must be creative, flexible, innovative, and always be on the move to be successful. Any new idea of a company is a surprise for their competitors. They must react and in return create surprising responses. Moreover, it is not only about the continuous improvement of the standard set of market drivers like time, money, and quality. It is also about finding entirely new ways to provide the same service to the customer or to create new demands at the customer site.

Now, a producer faces more competitors from different cultures with different ideas and with different cost structures. Imagine you are a successful producer of consumer cameras, and suddenly someone comes around and provides smartphones with a built-in camera of same or even higher quality.

Alternatively, you are a successful producer of navigation systems, and suddenly someone comes around and provides navigation and maps for free on the smartphone.

You know that these examples already happened. Similar market challenges can happen in your business, or you are the one who turns a market upside down.

Disruptive innovation happens with increasing frequency. The outcome is products that change the base architectures that are commonly agreed on the market. See chapter 4.6 about more details about the base architecture.

Although you can reach more people with your products by using global logistic networks and marketing channels, the markets are mature which leads to another phenomenon: the resurrection of craftsmanship. Section 3.6 covers engineering craftsmanship.

Another aspect of the globalization is global challenges of the world population. For example, a global challenge is the global climate warming, the potential impact of an asteroid, or the steadily increasing global population. Global challenges require global solutions: large engineered systems of systems that require collaboration between the systems and humans who develop and operate them.

We tend to focus on the technical aspect. More important is the human aspect. The technology is only an enabler. The way we use the technologies changes the world and not the technology itself.

The global challenges and the human's aspects are important ones. However, they would break the mold of the book, and I will not further detail them here. An interesting reading about this are the books *Limit to Growth* [Mea04] and *Collapse: How*

Societies Choose to Fail or Succeed [Dia11].

We cannot stop the globalization process unless we "uninvent" the internet, airplanes, and other global communication and transportation systems. Instead of arguing against it we must take the challenge and look for the opportunities.

3.2 Complexity and Dynamic

Many times in this book I mention the challenges of complexity and dynamics. The Oxford dictionary defines complexity as

A group or system of different things that are linked in a close or complicated way [Ox17].

In the following, I describe my understanding of these terms. I give two definitions of the term complex that cover two different aspects: an internal view of the characteristics of a complex entity, and an external view of how an observer sees a complex entity.

Definition Complexity [internal view]: *A characteristic of a complex entity is a large number of different elements and different kinds of relationships between the elements.*

The other definition specifies complexity from an external viewpoint:

Definition Complexity [external view]: *Complexity of an entity is a measure for the number of surprises from the perspective of an observer of the entity.*

In any case, it is hard to quantify the complexity. For example, you cannot unambiguously define a surprise and specify the exact number of surprises when an entity turns from simple to complex. It is a subjective valuation to define something as complex (see also figure 3.3).

Regarding the internal view definition, you can count the different element types and relationships. Although it is difficult to say at which number an entity becomes complex, you can use this approach to define your own categories in terms of values and assign appropriate consequences to the categories. For example, that an entity in the category *Complex* must get an additional verification and validation step, and an entity in category *Simple* not.

An exciting work is the First Law of Systems Engineering by Olivier de Weck [SW18]. The First Law of Systems Engineering is:

Given a fixed set of functional requirements R and human organizational architecture H, the total complexity of a system, C, is conserved. Complexity can be traded between its components and its interfaces and topology but cannot be decreased beyond a minimum level.

$$C - R - H = 0$$

whereas $C = C1 + C2 * C3$, and $C1$ stands for the number and heterogeneity of the components, $C2$ stands for the number and heterogeneity of the interactions between the components, and the scaling factor $C3$ stands for the dependency structure [Si13].

The First Law of Systems Engineering considers technical systems. However, the complex entity can also be an organization, or any other kind of something. Next, we have a look at organizations as an example of a complex entity.

An organization is often decomposed into smaller departments responsible and optimized for specific business functions. That leads to many different elements (=departments) organized in a hierarchical tree structure (figure 3.1).

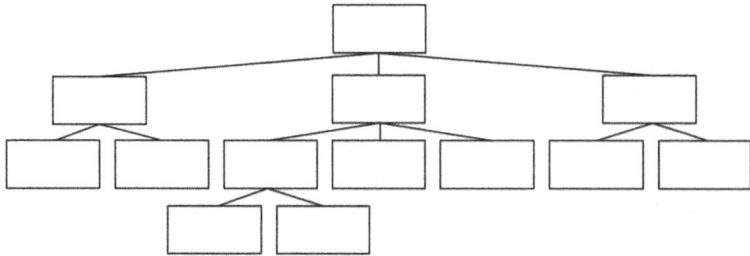

Figure 3.1: Tree Hierarchy: Typical organigram of an organization

As long as the organizational structure is a simple tree structure and the connections are almost identical with the primary communication paths it is not complex. There is only one kind of relationship. However, that is more theoretical, and a real organization has many communication paths of many different kinds - formal and informal - between the units on the same level or different levels, and not only along the decomposition relationships.

On the new engineering playfield, these paths are increasing in numbers since the new engineering game requires an increasing number of interdisciplinary tasks and pops up unexpected business cases that are not covered by the original structure of the organization (see section 3.3). Many different kinds of relationships and unofficial units come up and turn the organization into a complex system.

It is an exciting area of tension between simple and complex structures on different levels. Above a scenario is given how a simple structure can turn into a complex structure. The following is a scenario the other way round.

A single organizational unit is complex if it has many different relationships and consists of many different roles, i.e., it must perform many different business functions. One solution to reducing complexity may be to outsource some

functions into separate entities that reduce the complexity
of the entity while increasing the complexity of the entire
organization (figure 3.2) (e.g., [Luh87]). You can observe the
same concept for technical systems.

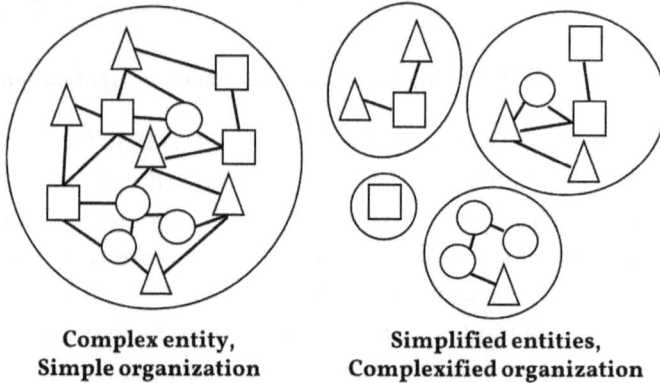

Complex entity,   **Simplified entities,**
Simple organization   **Complexified organization**

Figure 3.2: Levels of Complexity

You must differentiate between inherent and self-made complexity. Self-made complexity arises from poorly designed
structures and processes. By a redesign, you can reduce self-
made complexity.

Inherent complexity cannot be reduced. It mirrors the essential structure and processes necessary to perform the requested functions. However, as described above the complexity can be moved from one abstraction level to another.
That does not change the overall complexity but makes the
complexity more manageable.

The increased dynamics in the markets are in particular of
interest concerning complexity. The dynamic is one param-
eter for the frequency of surprises that happens in complex
systems. High dynamics lead to more surprises. That means

a simple system in a stable environment can turn into a complex system in a dynamic environment. The system does not change, but the environment.

The number of different elements and relationships that characterizes a complex system (*Definition Inner view*) is higher in a stable environment than in a dynamic environment (*Definition external view*).

Figure 3.3 depicts that relationship by combining the two complexity definition (internal and external) given above.

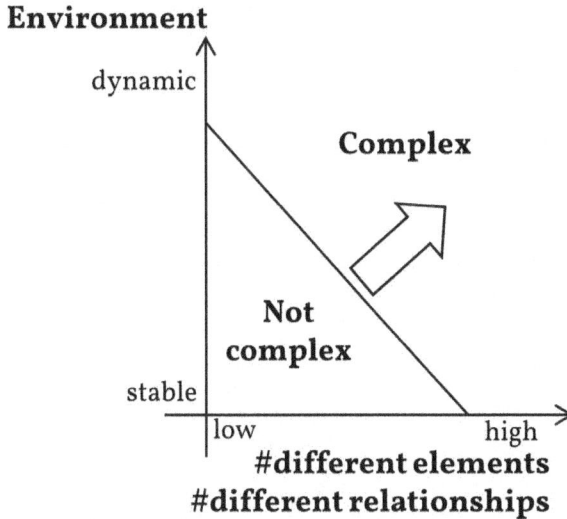

Figure 3.3: Relationship of Complexity and Dynamic

Strong process-oriented companies get in trouble in a dynamic environment. Processes cover known workflows and surprises are therefore not considered. The first question that should come up when a surprise occurs is not how to handle it, but who should do it [Woh04]. You must find the right person instead of the right process. Otherwise, the surprise is an exception in a process and deferred to another process where it is also an exception, and so on. Of course, you can argue that finding the right person to handle a surprise is also a kind of a process. However, that is a different level of acting.

Complex and dynamic products, organizations, and environments are the usual case in the new engineering game. It is not an exceptional situation. Players of the game must be prepared to deal with it in their daily business.

3.3 Conway's Law

Melvin Conway is a computer programmer who published his thesis - known as Conway's Law - in 1968:

Any organization that designs a system (defined broadly) will produce a design whose structure is a copy of the organization's communication structure. [Co68]

Think about the organizations and projects you have worked for. I guess you can recognize Conway's law in them. A product mirrors the structure of the developing organization because the organization structure constrains the solution space for the products accordingly. Vice versa the organization mirrors the product structure because the organization optimizes their work units by aligning them with the product structure.

You may have noticed that Conway said "communication structure" and not "organization structure". The organization structure is visible and often explicitly documented by organigrams. The communication structure is seldom documented and not well-known. Typically, there is a large overlap between organization and communication structure. In the following, I ignore the slight difference, but you should keep it in your mind.

Conway's law is highly remarkable, still true nowadays, and very important to be considered in both directions.

Think of a startup company that develops and sells an IoT product. They will explicitly or intuitively create an organizational structure that reflects the architecture of their product. It is a common optimization to put experts of the same field and who work on the same things close together in a single organizational unit. Figure 3.4 depicts the architecture of the product and the structure of the organization.

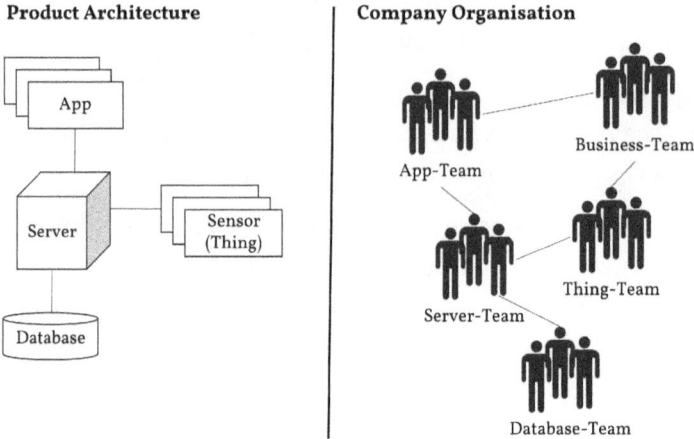

Figure 3.4: Conway's Law

Here, the organization has a component-oriented structure. In an agile environment, it would probably be a feature-oriented structure. In any case, the organization structure is strongly related to the product structure.

Future products of that organization will very likely always have an app, a database, and so on. From now on, the organization structure constrains the solution space of their products. Another kind of a product with a different base architecture (see section 4.6) would require a different organizational structure and can make existing structures like departments, roles, and people, superfluous. Naturally, the company
encounters some resilience against those product ideas. Besides that, it is unlikely that the company can achieve the idea of such a product at all.

If you plan to develop an innovative, maybe disruptive, product with an existing organization you typically face resistance. To avoid these obstacles, you can develop the product outside of the organization with a startup company. However, you need a merge strategy to bring the new product back into your organization after the development phase. Alternatively, you can keep it as a new organization on its own (see also chapter 4.1).

You should not conclude that the alignment of the organizational structure with the product structure is a bad thing.

James O. Coplien and Neil B. Harrison recommend that your organization structure is compatible with your product structure. Otherwise, you get trouble [Co04].

Conway's law is neither bad or good. It is just a fact about the relationship between the structure of the product and the developing organization. It is important to know that and to be aware of the outcomes:

1. If the structure of the organization does not reflect the structure of the product, the project gets in trouble.
2. It is hard for an organization to create innovations that do not reflect the existing structures.
3. A disruptive innovation requires organizational changes and must overcome organizational resistance.

3.4 Cyber-physical Systems

Beyond doubt, computer technology is one of the main drivers of progress and changes over the last decades. The PLCs played an essential role in the third industrial revolution (chapter 2.3). Now, the cyber-physical systems (CPS) play a vital role in the fourth industrial revolution (chapter 2.4).

Cyber-physical Systems (CPS) are not something that was invented by someone. They are an evolutionary step that was achieved by combining separate technologies. Simply put, they combine embedded systems with their sensors and actuators with digital networks like the Internet. CPS is a glue technology between the physical and the virtual world.

In 2008 Edward A. Lee defined CPS as [Le08]: *"Cyber-Physical Systems (CPS) are integrations of computation with physical processes. Embedded computers and networks monitor and control the physical processes, usually with feedback loops where physical processes affect computations and vice versa."*

The definition is around 10 years old and misses an aspect that is important nowadays: the connection of CPS with open networks like the internet. That connection closes the gap between the virtual and the physical world, and the openness enables the extension of the systems with functionalities that were not considered when they were developed.

CPS enables a lot of powerful applications, innovations, and new business models. Most of the technologies behind Industry 4.0, Digital Twins, and other trends of the new engineering game are based on CPS.

One of the challenges of CPS is the combination respectively collaboration between the organization units behind the embedded systems and internet technologies. Typically, these are separated organizational units with different kinds of people. Again, a scenario for interdisciplinary engineering (see section 3.5).

Besides the organizational challenges, there are several technical challenges. A CPS, for example, an autonomous automobile vehicle, is a complex system and operates in uncertain environments with lots of surprising events (see section 3.2).

The combination of organizational and technical challenges together with tough time-to-market, cost, and quality requirements is the real challenge of CPS.

The digital twin is a special kind of CPS that perfectly represents the merge of the virtual and the real world. One part of the digital twin exists in the real world. It is a physical product. The other part of the twin is a virtual representation of the physical product. They are interconnected and synchronize their values and states. You can change the properties of the physical product by applying the change to the virtual product. Vice versa the virtual product depicts the active status of the physical product.

The virtual part of the twin can cover only a few properties up to a nearly complete copy of the real-world twin part. A 100% coverage is impossible. Otherwise, the virtual part would be the same as the physical part.

An example of a digital twin is a smartphone app that controls some features of a car like the air conditioner. The app shows the status of the physical air conditioner in the car and the temperature. Another similar example is a smartphone app of a smart home system.

3.5 Interdisciplinary Engineering

The development of systems often requires profound software engineering capabilities as well as capabilities to engineer the physical thing itself, typically, electrical and mechanical engineering capabilities. Moreover, several specialty engineering disciplines like safety or domain-specific like medical, optical, or biological.

Although there is always a need for improvement, the specific engineering disciplines are not the weak spot on the new engineering playfield.

There was a time when each engineering discipline built their artifact, and finally, they were assembled to the complete system with some manageable interfaces between the artifacts. The increased complexity of the products and markets does not fit the strict separation of the engineering disciplines anymore. It requires a more holistic approach to master the systems and to provide the right solutions for the right problems.

1 plus 1 is not 2 anymore. 1 plus 1 equals 3, and the best solution is not the sum of the best parts, but well-balanced parts. The engineers of the parts can not do this task.

Many organizations are organized in disciplines, and a holistic approach is a challenge for them. Figure 3.5 depicts different levels of interdisciplinary collaboration.

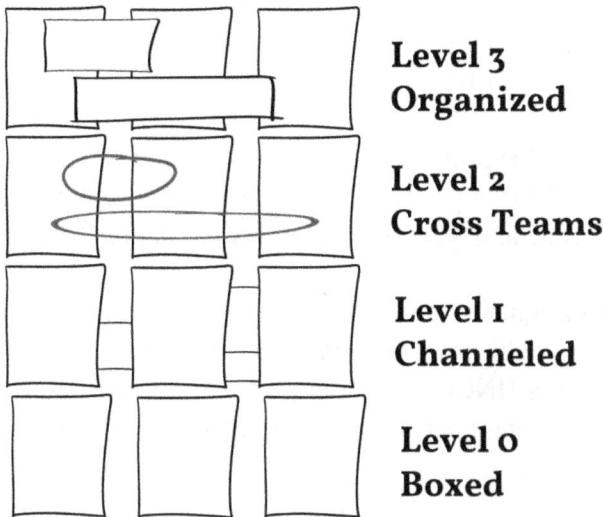

Level 3
Organized

Level 2
Cross Teams

Level 1
Channeled

Level 0
Boxed

Figure 3.5: Interdisciplinary Collaboration Model

There are four interdisciplinary collaboration levels:

- Level 0 - Boxed: There is no collaboration between the disciplines. A superior organization unit coordinates the work of the disciplines.
- Level 1 - Channeled: The disciplines collaborate via some predefined communication channels. For example, by exchanging requirements and architecture specification documents.
- Level 2 - Grouped: Interdisciplinary temporary working groups work on features of the product or service. When they finish the task, they break up.
- Level 3 - Organized: The organization structure contains permanent interdisciplinary units that are responsible for holistic tasks.

The interdisciplinary collaboration levels are not a maturity model. They are neutral, i.e., level 0 could be appropriate as well as level 3. The levels give orientation. For example, in combination with the Cynefin framework (chapter 4.2):

- Level 0 is probably ok for a simple context,
- Level 1 is appropriate for a complicated context, and
- Level 2 and 3 are for a complex or chaotic context.

Systems Engineering is *the* engineering discipline for interdisciplinary engineering. The *International Council on Systems Engineering* (INCOSE) is the global umbrella organization for systems engineering. They define systems engineering as follows [INC16]:

> *Systems Engineering is an interdisciplinary approach and means to enable the realization of successful systems. It focuses on defining customer needs and required functionality early in the development cycle, documenting requirements, then proceeding with design synthesis and system validation while considering the complete problem:*
>
> - *Operations*
> - *Cost & Schedule*
> - *Performance*
> - *Training & Support*
>
> *Systems Engineering integrates all the disciplines and specialty groups into a team effort forming a structured development process that proceeds from concept to production to operation. Systems Engineering considers both the business and the technical needs of all customers with the goal of providing a quality product that meets the user needs.*

According to this definition systems engineering perfectly satisfies the need for interdisciplinary engineering.

Like other engineering disciplines, Systems Engineering is a mature discipline in itself. It has its origin in the 1950ies. Huge projects during the times of the cold war like space, defense, or public telecommunication projects required more than the typical specific engineering capabilities.

In 1990 the National Council on systems engineering (NCOSE) was founded in the US to support research, education, and practical application of systems engineering. It was quickly realized that Systems Engineering is not only an emerging discipline in the US but worldwide, and NCOSE became INCOSE, the International Council on Systems Engineering.

Nowadays, systems engineering is a worldwide spread and recognized discipline. It is applied in many organizations, with systems engineering departments and systems engineering roles. You can get international accepted certificates or even study systems engineering at a university and receive a master degree. Several standards are available like the ISO 15288 about systems engineering processes [ISO15288] or the OMG Systems Modeling Language (SysML) [SysML17].

Typical areas of systems engineering, such as space, defense or aviation, already use systems engineering for many decades. As complexity and dynamics increase, more and more companies are adopting systems engineering to be able to master their systems and business. Also, their systems are not only the naturally large systems such as aircraft, ships, rockets, and so forth, but also physically small and complex systems such as hearing aids, control units, pumps, and so on. Bigness does not mean that it is also complex (see section 3.2). Moreover, to be small does not mean that it cannot be complex.

Although systems engineering is an interdisciplinary approach, it is often implemented within the meaning of the traditional silo thinking and is just another silo. This is a multidisciplinary and not a real interdisciplinary approach, as required by the INCOSE definition of Systems Engineering. Currently, INCOSE discusses to change the word "interdisciplinary" to "transdisciplinary" in the definition. The real tasks of systems engineering lie in the gaps between the disciplines and should not be encapsulated in a silo.

Figure 3.6 depicts the multidisciplinary scenario. There are clear borders and gaps between the disciplines and systems engineering is just another silo. That scenario conforms to the interdisciplinary collaboration levels 0 and 1 (see figure 3.5).

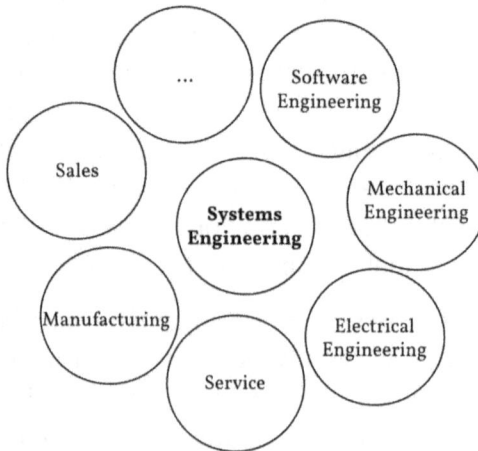

Figure 3.6: Multidisciplinary Systems Engineering

A real interdisciplinarity approach is depicted in figure 3.7. The disciplines have lots of overlaps (more than can be shown in a 2-dimensional figure). The interchange between the disciplines is not only about defined communication channels and document exchanges. The overlaps in figure 3.7 represent collaboration work in interdisciplinary teams and interdisciplinary roles performed by different or the same persons. That scenario conforms to the interdisciplinary collaboration levels 2 and 3 (see figure 3.5).

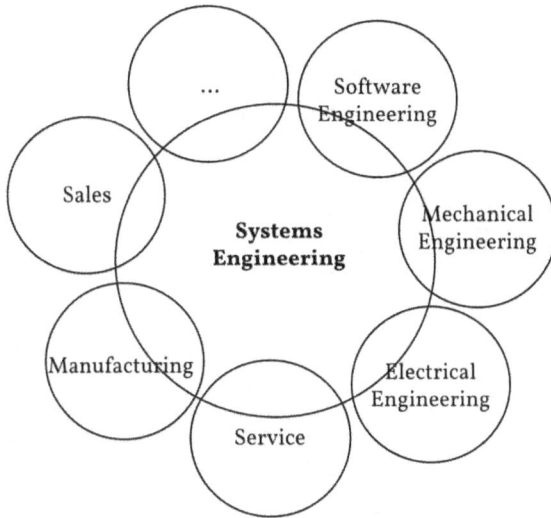

Figure 3.7: Real Systems Engineering without Gaps

The role of the systems engineer is to a large part being a moderator between the disciplines. Therefore, good soft skills is a mandatory capability of a systems engineer.

Chapter 4.9 presents a general-purpose and pragmatic methodology to elaborate the essential information on the system level in an interdisciplinary team.

Speaking of roles, did you also noticed the inflation of job names? That is another indicator of the change that there is no longer a clear separation of functions and responsibilities, for example, all the different manager kinds like facility manager or those people outside of my office who hand out parking fine notices calling themselves parking slot managers, or all the role names introduced by process frameworks like the Scrum roles Scrum Master and Product Owner or the Six Sigma roles like Master Black Belt or the Green Belts, and many many more.

The flood of role names mirrors the increasing complexity and the merge of engineering and business functions. If too many different functions are allocated to a role, it is natural that the role is split and new roles are born. For example, instead of just *Marketing* specialist, there are roles like a chief Twitter officer, chief blogging officer, or chief listening officer [Ec10]. These roles also point out another phenomenon: the inflation of chief roles that probably underlines the theses that their purpose is just to flatter employees without paying them more money [Gu12].

By the way, Kim Jong Il, the North Korean dictator, is probably the job title champion. He has more than a thousand titles including the Guiding Star of the 21st century [Gu12].

3.6 The Resurrection of Craftsmanship

Before the second industrial revolution (chapter 2.2) most products were created in manufactories by craftsmen. The productivity was low, but the production process was flexible and individual. The second industrial revolution changed the industrial landscape from craftsmanship to effective, but inflexible mass production.

The era of mass production survived the third industrial revolution. Nowadays we can observe a change. The markets again request flexibility, creativity, and individualism. Gerhard Wohland calls the overall pattern of the organizational complexity along a timeline the Taylor tube [Woh04]. Figure 3.8 shows the origin of the name *Taylor tube*.

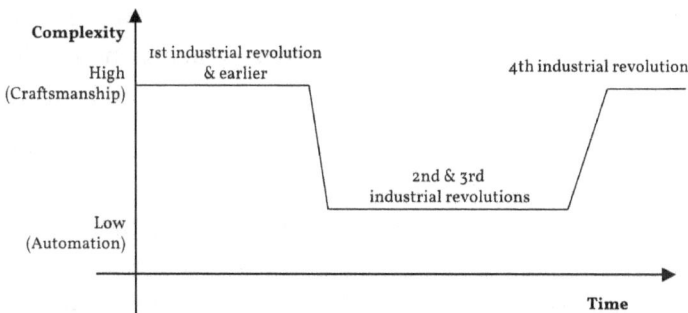

Figure 3.8: Taylor Tube based on Wohland

On the left side of the figure, you can see the times of the first industrial revolution. The downward sloping curve depicts the beginning of the mass customization era during the second industrial revolution. Machines performed more and more production steps. The production followed predefined fixed steps, and the complexity of the processes decreased.

The tightness of the markets led to the upward movement of the curve and heralded the beginning of the fourth industrial revolution. The markets request customized products. In combination with mass production, it requires complex automation and engineering craftsmanship.

Figure 2.9 in chapter 2.4 depicts the relationship between the complexity aspect and the industrial revolutions from another viewpoint.

To be still successful the focus must move from processes back to craftsmanship. Doubtless, processes have value. However, the value of craftsmanship is higher. It would be too simple if we think black and white here.

Individual customer requirements require that the customer is in close contact with the producer of the product. For example, agile methods integrate customers into the development process to create a tight loop between requirements and product feedback.

The Taylorism prevents this customer integration. The creation of the product is decoupled from the control and management. The latter ones typically also control the communication with the customer. Without being in touch with the markets and users, the right feeling for the right products can not emerge in the engineering departments. Moreover, the indirection costs time and quality. Both things you cannot cope with if you want to be successful.

3.7 New Work

New Work is a new movement that changes the working world. It mainly affects the knowledge workers.

One aspect is the mindset change of the younger work generation. The ongoing transition of an industrial society to a knowledge society in many countries changes the values of the people. Money, career, and status symbols are no longer the most important driving factors for them. More important is a good work-life balance, the meaning of life and meaningful work.

The internet is another enabler of the change. Nowadays, you can collaborate with people independently of their location.

That leads to questions like

- Why should someone else - called boss - tell me what to do if I can decide it much better and it would be much more effective?
- If I can work every time and everywhere, why do I have to be in the office from 9 to 5?
- Why must I record my working times? The outcome is important, not the work effort.
- If it is more important what I know than what I can do, what does that mean for my company if I quit my job?
- And many more.

Important values of the new work culture are independence, freedom, and co-design of the own work environment. These changes in our work culture must be in particular considered in an engineering organization. As stated in other chapters of this book the dynamics and complexity of the markets require flexibility, creativity, craftsmanship, and other capabilities that are closely related with happy employees.

The working culture that results from the previous industrial revolutions is entirely different. The separation of power and work introduced by the scientific management approach from Frederick Taylor led to the opposite of the new work values (chapter 2.2). Strong dependencies and constraints with little possibilities to influence the own work environment.

The New Work movement is another change challenge on the engineering playfield that must be considered by an organization. You can observe this change, for example, in agile methodologies like Scrum that promotes self-organized teams and a shift of control from the management (project leader, team leader) back to the engineers.

3.8 Gap of Slackness

Figure 3.9 shows the *Gap of Slackness*. The vertical axis represents the typical challenges for product vendors. They are, for example, more complexity, less time-to-market, less cost, and perfect quality. You can certainly list more of them for your domain.

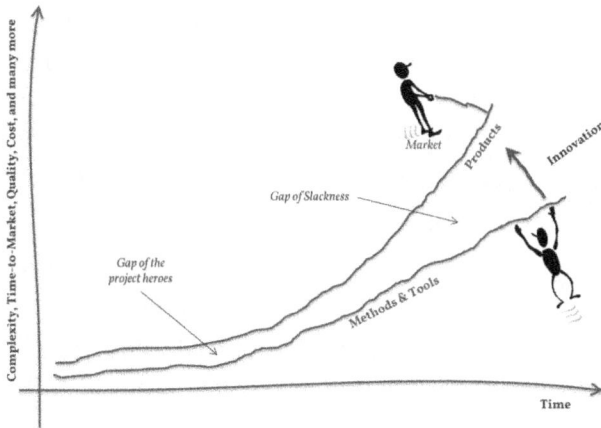

Figure 3.9: Gap of Slackness

The product curve along the time always increases and reflects the steadily increasing demand for improvements. The globalization and accompanying global dynamics change the linear slope of the curve to an exponential slope. More competitors with different backgrounds make the markets tougher for the vendors (see section 3.1).

We need methods and tools to develop those products, and they must also improve over time. The challenging factors are slightly different from the factors for the products. For simplicity, we borrow the same axis from the products for the methods and tools. The methods & tools curve is a little bit below the products curve. The distance is called the *Gap of the Project Heroes*. People who do outstanding work, save deadlines and have the right ideas at the right time (see also next section 3.9).

Even heroes have their limits. The exponentially increasing products curve leads to the *Gap of Slackness*. The organization does not spend time on sharpening their engineering tools because they are too busy developing products (with blunt tools). Until some point in time, the project heroes start to struggle. That overwhelms the project people, and you can recognize a change mainly on the social level. People are frantically, the atmosphere is getting unpleasant, and the sickness absence rate increases. On the engineering level, you observe the falling quality and missing deadlines.

The changes needed to raise the method and tool curve to close the gap require actions that are anything but easy. New products and product features requested by the markets can imply disruptive changes of existing product architectures. Conway's law (see section 3.3) says that this can lead to organizational changes. All in all that makes it much more than a technical challenge.

Critical is the point in time when you recognize the gap between the capability of your products and engineering tools. Actually, you must forecast the situation since the adoption of the engineering methods and tools cost time to be effective. On the contrary, the change costs time and money first. That's precisely what you do not need when it gets critical.

Chapter 4 presents principles, methods, and tools to close the *Gap of Slackness*.

3.9 Project Heroes and Burnout

Section 3.8 discusses the gap of the project heroes. Project heroes are the team members who make the impossible possible. However, their power is not boundless. If they get in trouble, it is an indicator that the organization gets in trouble.

Organizations should not rely on project heroes and be cleared of a charge by them. Project heroes can help out in short-term challenges, but should not be part of an organizational strategy. If the organization does not fill the gaps that were bridged by the project heroes, it implicitly continuously requests a high-speed working style.

Stress and work overload can lead to a burnout syndrome. The burnout phenomenon manifests itself in varying ways. It is not an easy task to detect it before it is too late, i.e., the only thing you will "see" from your project heroes are sick notes.

Although it seems that burnouts are a new phenomenon, it is pretty old. Typically, executives were affected [Le96]. Nowadays, they are still a typical target group, but also non-executives are "project heroes" and burnout candidates. In particular, in a new work environment, managerial functions and responsibilities are not restricted to executives (see section 3.7).

Nowadays, the information and communication technologies (ICT) allow a 24/7 reachability of employees and the time not thinking about work gets very little. A research from Connected Commons found out that most managers spend more than 85% of their time in meetings, e-mails, and other communication channels [Cr18]. The communication is crucial for the organizations and engineering projects, but it cannot just be added to the workload of the people. Communication and networks are valuable assets and must be managed. It is an organizational task to address that.

In addition to the organizational change, each must adapt his behavior to the new conditions of always-on communication. The article *Collaboration without Burnout* by Rob Cross et al. [Cr18] lists some best practices for your self-management how to face the challenge. They cover your beliefs, your role, schedule, and network, and your behavior.

There are many papers and studies around that analyzed the relationship between burnouts and ICTs, for example [Ko16].

You need a careful look at the commitments of your team members. Is it based on enthusiasm or closing gaps of deficits? Is it temporary or permanent?

4. Tools for the New Engineering Game

There is no silver bullet how to master the challenges sketched in the previous chapters. The best strategy highly depends on your specific business and markets. It is strongly recommended to have a well-filled toolbox with principles, patterns, methods, and more to have the right action at hand.

Processes do not work well on the new playfield. They can only cover known events and scenarios. However, now the unknown is the order of the day.

This chapter provides a proper toolbox for the engineering organization and people to master the challenges of the digitalization.

The first part covers more organizational aspects while the second part of this chapter focuses on tools for engineering tasks.

In the first section 4.1, you learn about the high-level strategy *Turtle & Rabbit of Conway's Law* (see also chapter 3.3).

Section 4.2 presents the Cynefin Framework to guide actions in complex environments.

The Business Model Canvas presented in the next section 4.3 supported by the Value Proposition Design approach presented in the following section is a great tool to elaborate and describe a business model. In section 4.4 we have a brief look at the Business Model Navigator approach. Finally, you learn in section 4.5 the Business Motivation Model that defines the terms vision, mission, strategy, objectives, and goals, as well as the relationships between them.

The second engineering-focused part starts with the concept of the base architecture in section 4.6. It is a helpful tool to fasten the starting point of product development.

The zigzag pattern in section 4.7 clarifies the relationship between architecture and requirements on different levels (customer, system, component, and so forth).

Section 4.8 introduces Design Thinking. It is a customer-oriented methodology to solve problems. It is in particular helpful to find solutions beyond existing base architecture in the organization.

The following section 4.9 provides a pragmatic, general-purpose methodology for product development. An architecture kind independent of technology is the functional architecture described in section 4.10.

A product development creates lots of complex data about requirements, architectures, and so forth. To access the data from different viewpoints, it must be stored in models and not in documents. Section 4.11 describes the Model-Based Engineering approach.

A crucial part of an engineering project is to master the requirements. The REThink 4.0 approach presented in section 4.12 disrupts the fundamental concept of text-based requirements engineering.

The final section 4.13 strengthen the importance to perform the engineering processes agile and lean.

4.1 Conway's Turtles and Rabbits

Turtles and Rabbits is an organizational pattern to deal with the challenges of Conway's Law presented in section 3.3. A disruptive idea for a new product, i.e., a fundamental change of the base architecture (section 4.6), typically lead to resistance by the organization. For example, because parts of the organization get superfluous for the development or production of the new product.

It cost effort and in particular, time to overcome the resistance with the additional risk that the development of the new product is not effective and powerful. The established organization is like a turtle: consistent, stable, and slow. That is very useful since the established business must go on. However, a turtle organization is not the best to create disruptive new products. The organization is resistant to changes with regard to fundamental changes in the basic architecture. It does not mean that the organization is not innovative.

On the other side^, a rabbit is fast and can continuously change the direction. It is hard for a rabbit to be a friend of a turtle. A rabbit organization is open to fundamental changes and can quickly adopt new methods and implement disruptive product ideas.

Figure 4.1 summarizes the characteristics of the two work styles.

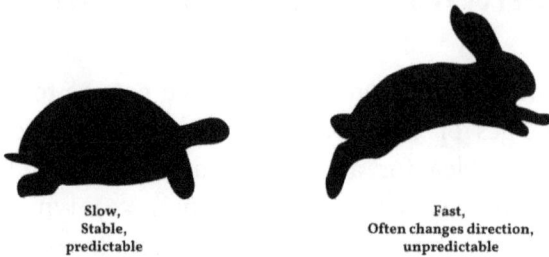

Slow,   Fast,
Stable,   Often changes direction,
predictable   unpredictable

Figure 4.1: Conway's Turtles and Rabbits

The rabbit is an explorative work style that deals with new problems and challenges. Prototypes and experiments test new solutions. The work is characterized by uncertainty. Mapped to the Cynefin framework (section 3.2) the work is mostly located in the *Complex* section.

The turtle work style is characterized by predictability. The tasks are well known and can be described by processes. This part keeps care of the daily business. Mapped to the Cynefin framework (section 3.2) the work is mostly located in the *Complicated* section.

In order not to stand in each other's way, you explicitly address the two different ways of working and cultures.

The Bimodal IT approach proposed by Gartner is also based on the pattern [Gar18]. However, Bimodal IT is also a very controversial approach. One point of criticism is that you should enable the old IT to work more exploratively instead of building another organization in parallel. The approach can help to avoid the difficulties of changing and modernizing an organization. But the "old" organization also has to adapt and change. For example, the isolated view of IT does not fit the requirements of digital transformation, where the processes of a company must be considered end-to-end starting at the customer.

The Turtles and Rabbits pattern is also called Greenfield versus Brownfield, legacy versus emergent, or just old versus new engineering. Whatever you call it, there is a broad agreement that innovation needs its own space. So, the Turtles and Rabbits pattern is also not a recommendation for bimodal IT. Rather, it is just an observation that there are these two different forms and the recommendation to address that explicitly. Bimodal can be one possible solution.

The two different work styles typically require different kinds of people. Not everyone like and can work with uncertainty. Vice versa people who like uncertainty and explorations get easily be bored by a predicted work style. Of course, some people can work with both styles.

Be careful to establish a fair culture to avoid getting a welcome and unwelcome work area. Both - the turtles and the rabbits - are valuable, relevant, and good. Typically, the disruptive product is more in the spotlight than the standard parts, and the rabbits could easily be seen as the stars of the organization.

You can separate the rabbits and turtles in different roles, teams, departments, or even different legal organizations. In the following, the *organization* means any of these options.

The development of the BMW electric vehicles i3 and i8 are an excellent example for the rabbit and turtle pattern. Norbert Reithofer (at the time CEO of BMW) said in an interview in 2014 [Re14],

> For big innovation leaps, you have to leave the usual paths. Otherwise, the development of the i3 would not have been possible which sets new

standards regarding technology and lightweight construction. That's why we have let work the development team for the BMW i3 outside the typical corporate structures. Such a step makes sense in such a case for some time.

The organization BMW has a strong focus on combustion engines, although the company strategy may have changed. The development of an electric vehicle is a strike against the base architecture and the related manifested organization structure.

At some point in time, the disruptive product is not disruptive anymore. It is established in the organization and the market and fits more to the turtle than to the rabbit organization. You need a strategy on how to deal with that situation.

First, define beforehand when do you expect the change from the rabbit to the turtle working style. It can be a specific time like the start of production, or the start of the sale, or after being one year on the market. In any case, it should be planned and not a reaction to unpleasant surprises.

The following step is not an easy one. If you transfer the product to the turtle organization, what happens to the rabbit organization? You have the following options:

1. Close the rabbit organization and discharge the rabbits.
2. Close the rabbit organization and transfer the rabbits to the turtle organization.
3. Find a new project for the rabbits.

Option 3 seems to be the best one. However, it often could not be appropriate for an organization to do more innovations, because there are not enough resources to spend the effort. For example, a startup company with 20 people that has placed its first product on the market. They need all their resources to produce, sale and maintain the existing product.

BMW has picked option 2. In the interview mentioned above, the CEO of BMW, Norbert Reithofer, also said:

> However, if you need more advanced ideas, it makes sense to separate teams. They should only be brought back into the organization at the right time.

It seems that BMW was not very successful with option 2. In the news, you can find reports that top experts of the electric vehicle team from BMW were recruited by Chinese competitors and left BMW during that time.

Good moves on the Engineering Game Board

- Check, if you have rabbit and turtle working styles.
- Explicitly address the difference between rabbits and turtles.
- Keep the distance between rabbits and turtles as low as possible.
- Plan the end of the rabbit working style on time.
- Promote a fair culture. Rabbits and turtles are worth the same.

4.2 Cynefin Framework

Finally, I would like to briefly present the Cynefin framework by Dave Snowden [Sn07]. It is a clear and concise model to describe the domains complex, complicated, chaos, simple, and disorder.

The idea of the Cynefin framework is to provide a tool for decision-makers to see things from different viewpoints.

Figure 4.2 depicts the Cynefin framework with the five domains: simple, complicated, complex, chaotic, and disorder.

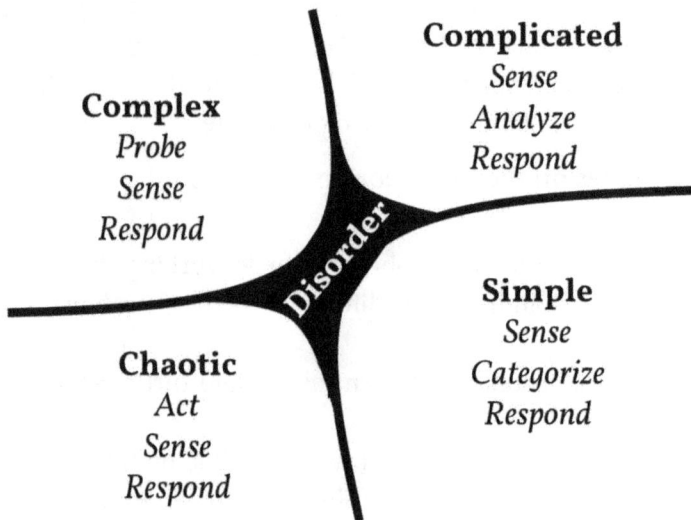

Complicated
Sense
Analyze
Respond

Complex
Probe
Sense
Respond

Disorder

Simple
Sense
Categorize
Respond

Chaotic
Act
Sense
Respond

Figure 4.2: Cynefin Framework

The simple domain hosts the "known knowns", i.e., things we know that we know them. The reaction on events in the simple domain is to determine the facts ("sense"), to categorize them, and finally to act based on rules or best practices of that category.

An example is a support request by a customer. First you collect all available information, then you categorize the request as a first, second, or third level support request, and finally, you act accordingly.

The simple domain is the location for clearly defined rules and legal structures.

The complicated domain is similar to the simple domain except that you cannot derive an action by just applying a rule or practice from the assigned category. It requires an analysis to find the best action. Typically, there are more right answers, and an expert must decide on the best one.

The complicated domain hosts the "known unknowns" and is the primary domain of the engineers. Most technical systems are located in the complicated domain. For example, an airplane can be understood by engineers even if it is sophisticated (= complicated).

The complex domain leaves the predictable world. The cause-effect-relationships can only if ever, be understood afterward. Therefore the recommendation is first to probe the situation, then sense it, and finally respond. Cyber-physical systems can be complex (see section 3.4). Their openness makes them partly unpredictable.

The chaotic domain is without any cause-effect-relationship. The best you can do is to act first to stabilize the situation. Then sense it and respond to turn the situation from chaotic to complex.

The September 11 attacks are often mentioned as an example of a chaotic scenario. Acting had the highest priority after the attacks. Sensing and responding came later.

The exclusive domain *Disorder* in the middle of the Cynefin framework represents a situation where there is no clarity about the situation. It is not chaotic. In the chaotic domain, you at least know that it is chaotic. You do not know that in the disorder domain. It is hard to decide if a disorder situation exists. If you can categorize the situation, it is not a disorder.

Good Moves on the Engineering Game Board

The Cynefin framework is a valuable tool to guide the course of your actions in a dynamic environment. Be aware of the different domains and act according to them.

4.3 Business Model and Value Proposition Canvas

The Business Model Canvas (BMC) is a one-page visual chart for a business model. Alex Osterwalder et al. have presented the canvas in their book *Business Model Generation* [Os10].

The canvas is a powerful tool to elaborate a high-level business model, i.e., the main pillars of a business like resources, customers, value proposition, and finance. It is a chart that shows all this information on a single page.

Figure 4.3 depicts the Business Model Canvas template. The numbers are added to reference the separate sections in the following descriptions.

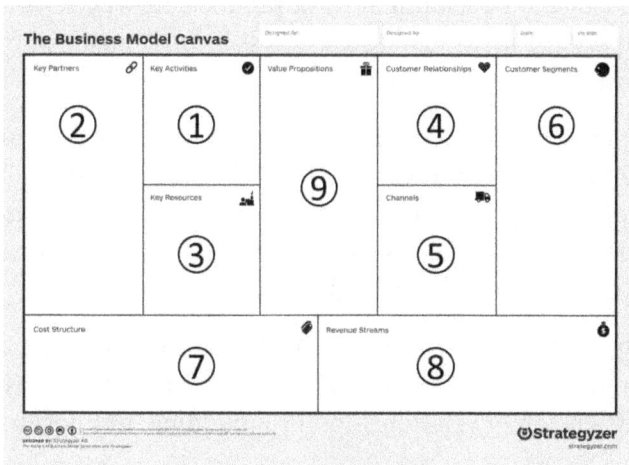

Figure 4.3: Business Model Canvas Template (Source: strategyzer.com)

The left side of the BMC covers the resources of the business. The right side covers the customer aspect, and the bottom depicts the financial fundament. The middle section of the canvas is about the value proposition.

Next, we have a more detailed look on the sections starting with the resources (#1 to #3) on the left side.

The *Key Activities* section (#1) lists the most important activities that are necessary to offer the value proposition. For example, building a community in social networks, or developing a supply chain that satisfies tough cost and time requirements.

Later, you should map the activities to business units and roles. A helpful tool for that step is the functional architecture. Section 4.10 describes the functional architecture in the context of the development of a product. However, you can also apply the method at the business level. A business is also a special kind of a complex system.

The *Key Partners* section (#2) lists external organizations and persons who are essential for the business model. For example, a supplier or a service provider. This section also includes networks like supplier networks, i.e., a particular focus on relationships instead of the partners itself.

Finally, the *Key Resources* section (#3) contains the most required resources like knowledge of the employees, patents, money, customer relationships, or a building, or a location.

The right side of the BMC depicts the customer view (#4 to #6).

The section *Customer Relationships* (#4) describes which relationships are expected by the customers, and which relationships are planned or already established by the business, for example, personal on-site support, online support for free, self-service, automated services, or communities.

The section *Channels* (#5) covers the communication with the customers. A business can directly communicate with its customers through online portals or on-site stores. It is also possible to have an indirect communication with the final customers, for example, via distributors.

The *Customer Segments* (#6) lists the customers itself. The segments cluster customers with similar needs and attributes. For example, mass market, males, females, older people, academic people, and so forth.

The bottom of the canvas depicts the financial fundament of the business. The *Cost Structure* section (#7) lists the central cost positions of the business model. You should distinguish between fixed costs that are independent of the number of sales like salaries and variable costs like supplies.

Part of the cost structure is also the cost strategy. You can focus on minimizing all your costs, for example, a discounter, or you can focus on increasing the value of your products or services, for example, a vendor of luxury goods.

The *Revenue Stream* section (#8) describes how the business makes income. There are several ways to generate a revenue stream. The customer segments or channels can also distinguish it.

The middle of the canvas depicts the central aspect of a business model: the *Value Propositions* (#9), i.e., the products and services offered by the business and the value for the customers. They distinguish the business from its competitors. Besides the provided functionality by the product or services, the quality is a crucial part of the value proposition. Quality attributes are, for example, performance, price, accessibility, or newness.

The value proposition is the primary aspect of a business model, and everything else is only required to make the value proposition real. Alex Osterwalder et al. have published another book only about the value proposition. Their book *Value Proposition Design* was published four years after the book *Business Model Generation* in 2014 [Os14]. Again he proposed a canvas to elaborate the value proposition.

Figure 4.4 depicts the Value Proposition Canvas (VPC). I have added numbers again to reference the sections in the text. Besides the value proposition, it is also linked to the *Customer segments* section of the BMC. The top of the canvas shows the relationships to BMC sections.

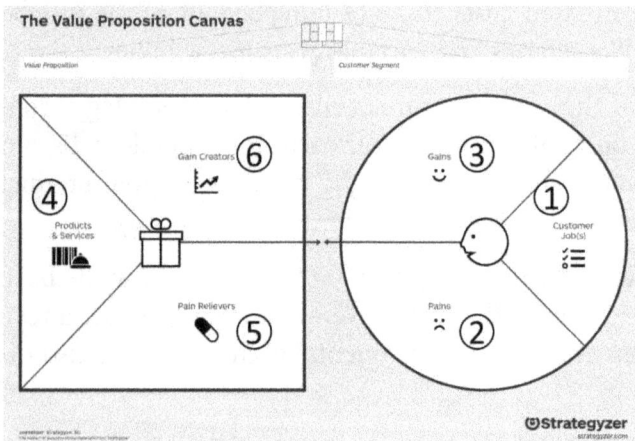

Figure 4.4: Value Proposition Canvas Template (Source: strategyzer.com)

The customer section in the value proposition canvas has three sub-sections:

The *Customer Jobs* (#1) is a list of tasks and challenges of the customers. Of course only in the context of the targeted business. Let us assume you want to elaborate the value proposition for an online meeting business. The customer jobs in this context could be:

- Join a meeting
- Connect audio
- Share webcam stream
- Present something
- Discuss something
- Leave meeting

The list of customer jobs is similar to the list of use cases elaborated with the methodology presented in section 4.9. However, the customer jobs typically cover the broader field of the business while the use cases are specific for a product or service.

The *Pains* (#2) section lists the problems and challenges of the customers when doing their jobs. For the online meeting business the pains could be:

- Audio does not work
- Wrong application shared
- Too many people talking at the same time
- The webcam is on without knowing it

The *Gains* (#3) section covers the positive side of the customer jobs. The values and expectations of the customers doing their jobs are listed here. For example, for the online meeting business

- Easy connect with remote people
- Reduce travel cost and effort
- Enables remote work
- Enables international teams

The left side of the VPC also has three sections as counterparts to the three customer fields on the right side.

The *Products & Services* (#4) section lists the offer of the business for the customer jobs. For example, an online meeting platform with screen sharing, webcam, and audio by voice over IP as well as telephone. I know that this is not very innovative or disruptive, but sufficient for our example. Moreover, it is probably not a good idea to disclose a promising idea in a book.

The *Pain Relievers* (#5) are parts of the products and services that help the customer to overcome the pains. For the online meeting platform pain relievers could be:

- "Mute all" feature
- Audience view depicted in a separate window
- Single talk audio filter

The *Gain Creators* (#6) are solutions that satisfy the gains of the customers. Gain creators of the online meeting platform are, for example:

- 1-Click access feature
- Team office feature

The VPC is no rocket science. It is simple but powerful. Among others because it is simple.

It can be combined with the BMC and worked out in the same workshop. However, it can also be used independently of the BMC.

With the value proposition, we have covered the complete BMC.

The canvas omits one important point: the objectives of the business. Do you want to earn as much money as possible, or do you want to save the world? Whatever it is, it is the motivation behind your business, forms the culture of your organization, and affects all your decisions and actions. The objectives of product development are covered in section 4.9 as part of a product development methodology. Section 4.5 explains how objectives relate to mission, vision, and strategies.

The BMC and VPC are creative tools for people to design a business model in workshop sessions jointly. A particular workshop setting can be: Use a whiteboard or a pinboard, post-it notes, and board markers. Discuss and fill-out the sections of the BMC.

Willibald Erhart describes another approach where each member of the group works for 20 minutes on a single section of the canvas on its own [Er17]. Then they rotate and work on another section that was previously worked out by another group member, and so on. The ideas of the workshop members are reviewed and revised several times to be finally mature and known by all participants. Rotate until everyone has worked on all sections. Although it is very valuable to do that in a single room to have face to face communication, you can apply that approach also remotely and also over a longer period.

This method is a variation of the 6-3-5 brainwriting method [Ro69]. 6 people write down 3 ideas in 5 minutes on a worksheet. After that, they swap around the worksheets, and after 6 rounds every worksheet was processed by every participant.

If you agree on the elaborated business model, you must detail it, derive a business architecture, and do further actions to build a business that implements the business model. How to do that is out of the scope of this book. A starting point can be the methodology to develop complex systems in section 4.9 that can also be applied to businesses. As mentioned above business is a special kind of a system.

Good Moves on the Engineering Game Board

Create a Business Model Canvas of your business and a Value Proposition Canvas of your value proposition in a workshop together with your colleagues. It costs not much effort and has a high value. Simplicity wins!

4.4 Business Model Navigator

The Business Model Navigator (BMN) developed by the University of St. Gallen is a method to elaborate a business model ([Ga18], [Ga15]). Part of the navigator is a list of business model patterns that capture most of the existing business models.

There is no collective agreement what is part of a business model. The BMC with 9 different section (see previous section 4.3) also covers only the parts that the authors think are important for a business model.

A BMN business model pattern has only four dimensions: the Who, the What, the How, and the Value (figure 4.5).

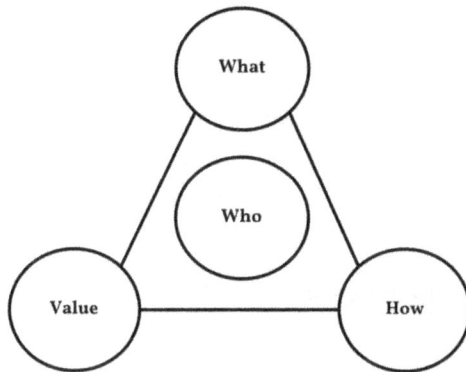

Figure 4.5: Magic Triangle (Source: [[Ga18](#BibGa18)])

Similar to the BMC the BMN impresses with simplicity, and at the same time, it is powerful enough to provide a clear picture of the business model for the stakeholders.

The **Who** dimension lists the customer of the business. In the BMC, the "Customer Segments" section covers the *Who* dimension. The customer plays a central role in the BMN.

The **What** dimension covers the offering of the business, i.e., the products and services, and the value for the customer. It is similar to the "Value Propositions" section of the BMC.

The **How** dimension describes how the business creates and offers the value proposition. In the BMC, the "How" is covered on the left side by the "Key Activities" section.

The **Value** dimension covers the finances of the business, i.e., the cost structure and the revenue stream. Alternatively, simply said, it answers the question of how to make money with the business. In the BMC, the sections *Revenue Stream* and *Cost Structure* cover the financial aspects.

A basic idea of the BMN is that innovation is based mainly on the recombination, transformation, and so forth, of existing things. The team behind BMN has analyzed many business models and finally came up with 55 business model patterns.

For example, the pattern *ADD-ON* affects the *What* and *Value* dimensions. The core product or service is priced competitively, and add-ons achieve the revenue. Ryan Air is a typical example of an ADD-ON business model.

The BMN methodology describes three core steps:

1. Analyze the context of the business to understand the trend of the market.
2. Select business model patterns that match the analysis outcome from step 1.
3. Reality check of the selected business model patterns.

You find a detailed description of the methodology in [Ga15].

The University of St. Gallen provides several materials for a BMN workshop like worksheets and cards at www.bmi-lab.ch.

Good moves on the engineering playfield

- The BMN is an alternative or addition to the BMC.
- Find the business pattern that matches your business and analyze how well it fits the market.

4.5 Business Motivation Model

Do you know the difference between objective and goal? Or between strategy and tactic? Or between mission and vision? The Business Motivation Model (BMM) defines these terms and the relationships between them [BMM15].

The BMM provides a structure and definition of high-level business terms. With BMM you can describe the motivation of your business (ends) and the means how to achieve these ends.

The content in this chapter is inspired by the description of the BMM in the book *OCEB 2 Certification Guide: Business Process Management - Fundamental Level" by Tim Weilkiens et al. [We16b]

It is a world-wide standard published by the Object Management Group (OMG). So to say, it is a common understanding of business terms. That is helpful to avoid misunderstandings and grueling discussions, and it gives you a structure to work out a clear picture of your business. It is not an alternative to the BMC and VPC (section 4.3). It is an addition that covers other aspects and has a different focus.

The BMM standard provides semantic and the data structure of business terms, but no notation. It is up to you, how to show the information: by pure text, or self-defined symbols, or a profile of the Unified Modeling Language (UML) [UML17], and so forth.

Figure 4.6 depicts the main pillars of the BMM.

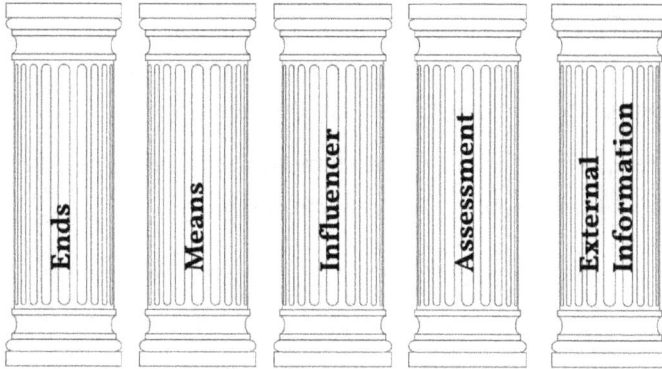

Figure 4.6: Pillars of BMM

The *Ends* pillar describes the vision and derived objectives and goals of the business.

The *Means* pillar covers the main strategies and tactics to achieve the ends.

The *Influencer* section describes external impacts like competitors or market trends and internal impacts like IT infrastructure.

The *Assessment* evaluates the impact of the influencers on the ends and means of the business.

The *External information* pillar covers further information about the business that is not covered by the BMM vocabulary like business processes.

In the following, we have a more detailed look on each BMM pillar.

We use the concept model approach to describe the BMM terms. It is similar to the original definition of the BMM in the official specification document [BMM15], but a bit simplified for a more natural understanding.

A concept model - also known as a domain-knowledge model [We16a] - is a model and graphical representation of the concepts and relationships between them.

Figure 4.7 shows the concept element itself and three different kinds of relationships in a concept model.

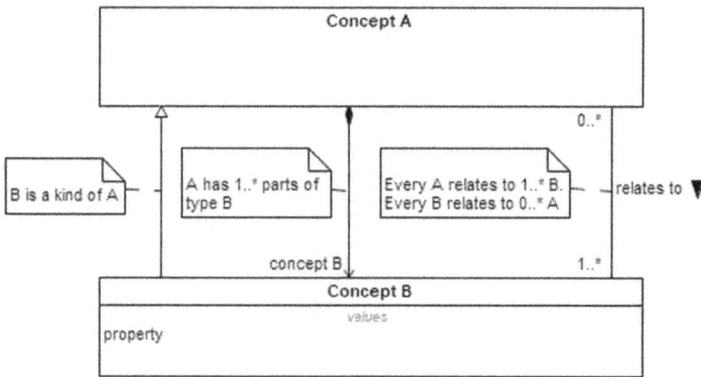

Figure 4.7: Concept Model Concepts

Here, the concept model is created with the Systems Modeling Language (SysML) [SysML17] in a modeling tool. However, you can also use an office or drawing application, or even just a pen and piece of paper to create a concept model.

A concept is depicted by a box with the name of the concept inside the box, and an additional compartment provides a list of properties of the concept. For example, in figure 4.8, the concept *Customer* has three properties: *segment, kind,* and *priority*.

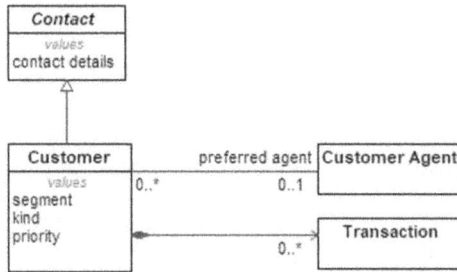

Figure 4.8: Concept Model example

The *"is a kind of"* relationship between *Customer* and *Contact* defines a concept as a specialization of another concept. The *Customer* is a special kind of a *Contact*. It has the same semantic and properties, i.e., a *Customer* is a *Contact* and has *contact details*. Additionally, it has more constraining semantics. A *Customer* is not any *Contact*. It is a *Customer* of the business with an active contract, and a *Customer* has three more additional properties, i.e., in summary, it has 4 properties: *contact details, segment, kind,* and *priority*.

The concept *Contact* is an abstract concept. It is a general concept that must be specialized by concrete ones. An abstract concept is depicted by italic letters. The target of a "is a kind of" relationship must not be abstract and can also be a concrete concept.

The part relationship between *Customer* and *Transaction* in figure 4.8 defines part properties of a concept that are defined by another concept. The black diamond symbol depicts the side of the owner. The other end has an arrowhead, a multiplicity that defines the minimum and maximum amount of the parts, and an optional role name.

In figure 4.8 a *Customer* can have zero to many (symbol *)
transaction parts defined by the concept *Transaction.* The part
relationship has no role name.

The relationship between *Customer* and *Customer Agent* in
figure 4.8 describes the role of a concept in relation to another
concept. A multiplicity and an optional role name stand at
each end of the relationship line. You read the relationship
in one direction as "Every <concept A> has <multiplicity B>
<concept B>s in the role of <role name B>.".

In figure 4.8, you can read "Each *Customer* has zero to one
Customer Agents in the role of *preferred agent*". Sounds a
little bit bumpy, but hits the point. And in the other direction:
"Each *Customer Agent* has zero to many *Customers.*".

Back to the BMM, the following figures show the definition of
the structure of BMM business terms by using the conceptual
model approach.

Figure 4.9 depicts the *Ends* elements of BMM.

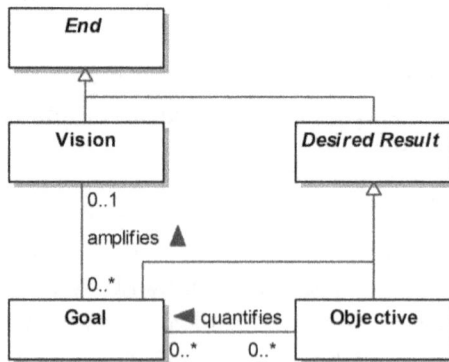

Figure 4.9: BMM Ends

End is the generic abstract concept representing the purpose of the business. The *Vision* is a special concrete *End*. It describes a desired, potentially unreachable state of the business in the future.

To give an example, I describe a business motivation model of the publishing company MBSE4U. Although MBSE4U is real (you have a book of MBSE4U in your hands or a digital version on your device), the business motivation model is pure fiction.

The (fictional) vision of MBSE4U is *"MBSE4U is the worldwide leading knowledge provider for model-based systems engineering."*

You may know other usages of the term *Vision* that describes the desired state outside of the business instead of a state inside the business. For example,

- *"Our vision is a world where everyone can be connected."* – Nokia
- *"A PC on every desk in every home."* – Microsoft

You can call the Nokia and Microsoft statement a vision. However, in the sense of BMM, it is not a vision. It must be a state of the business itself.

It seems that Ford has thoroughly read the BMM specification. The vision of Ford is

"Become the world's leading consumer company for automotive products and services." – Ford

Well done!

In figure 4.9 *Goals* and *Objectives* are special *Desired Results*, and the *Desired Result* is a special *End*. In practice, the terms *Goal* and *Objective* are often used synonymously. In BMM, they are different concepts.

A *Goal* is directly related to the *Vision*. It is a long-term state that the business must achieve to amplify the *Vision*. For example, MBSE4U has the goal:

"MBSE4U books are among the most important references in model-based systems engineering."

That is a nice goal, but how do you know when you have achieved the goal. *Objectives* quantify the *Goals* and make them measurable.

An objective is something you can measure. The following is an objective that quantifies the goal from our MBSE4U example given above:

At least one MBSE4U book is on the Amazon TOP10 list of engineering books.

Typically, more than one objective quantifies a goal.

Figure 4.10 depicts the *Means* elements of the BMM.

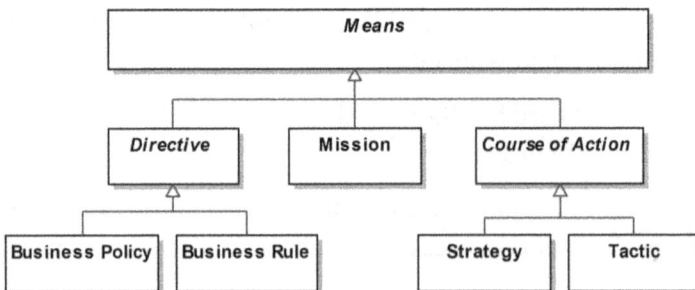

Figure 4.10: BMM Means

The *Means* describe the actions necessary to achieve all the ends. Or in the words of Joel Barker[1]

"Vision without action is a dream. Action without vision is simply passing the time. Action with vision is making a positive difference."

The BMM *Means* include three main concepts:

- the *Mission* as a counterpart to the *Vision*,
- the *Course of Action* as a counterpart to the *Desired Result*,
- and the *Directives*.

The BMM separates the ends and means and applies herewith the separation of concerns principle. It is a common principle in engineering for requirements management and product architecting. You can have a great (stable) vision and change the means over time to achieve it.

The *Mission* describes the ongoing activity of the business to make the *Vision* a reality (figure 4.11).

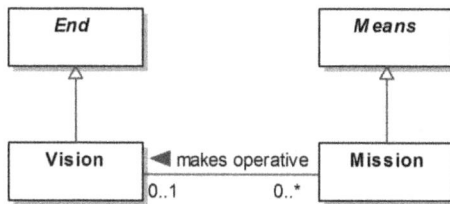

Figure 4.11: BMM Vision and Mission

---

[1]Joel Barker is a futurist, authors, lecturer, and filmmaker about leadership and business (www.joelbarker.com/bio/).

A *Mission* statement has three parts:

- an action,
- a product or service,
- and a customer or market.

The mission statement of MBSE4U is

Publishing books and supporting material about systems engineering worldwide on the pulse of the markets.

A *Course of Action*, i.e., *Strategy* or *Tactic*, supports the achievement of the *Desired Result*, i.e., *Goal* or *Objective*.

Strategy and *Tactic* are special *Course of Actions* and the counterparts of the BMM *Goals* and *Objectives* (figure 4.12).

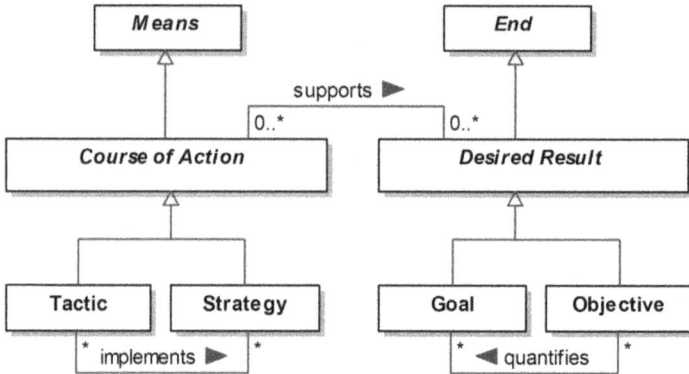

Figure 4.12: BMM Course of Action

As with *Goal* and *Objective*, the terms *Strategy* and *Tactic* are often used synonymously, but BMM again differentiates between the concepts.

Figure 4.12 shows that a *Tactic* implements many *Strategies*. They are related with *Goal* and *Objective* via the generic term *Course of Action* that supports the *Desired Result*.

For our example, strategies to support the goal *"MBSE4U books are among the most important references in model-based systems engineering."* are:

- *"We publish books of authors who are well-known in the systems engineering community."*
- *"We publish books about new relevant topics."*

These strategies give the business direction towards the goal that amplifies the vision *"MBSE4U is the worldwide leading knowledge provider for model-based systems engineering."*

For our strategy *"We publish books of authors who are well-known in the systems engineering community.",* tactics are:

- *"We get in regular contact with promising authors."*
- *"We visit systems engineering conferences to get in contact with well-known potential authors."*

The third part of the *Means* section are the *Directives*. They support *Desired Results*, i.e., *Goals* and *Objectives*, and governs *Course of Actions* (figure 4.13). Concrete *Directives* are *Business Policies* and *Business Rules*.

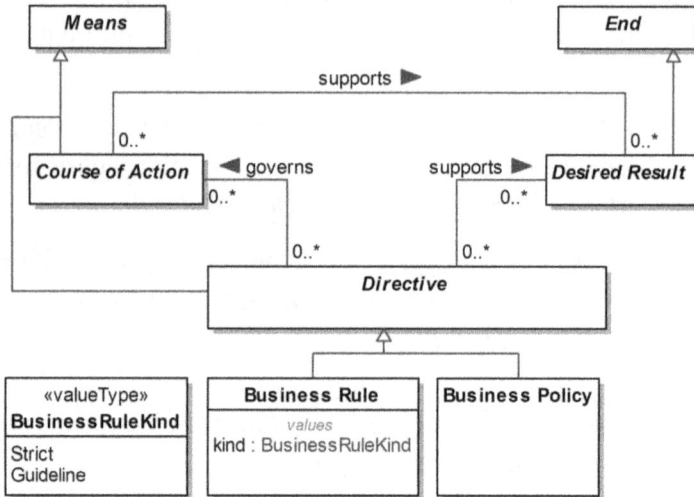

Figure 4.13: BMM Directives

A *Business Rule* has an enforcement level that can have one
of the two values:

- *Strict* - The business rule is mandatory, for example, a
  labor-law related instruction.
- *Guideline* - The business rule should be obeyed, but
  exceptions are allowed.

MBSE4U business rules are for example:

- *Enforcement level "Guideline": A booklet has 80-150
  pages, and a book more than 150 pages.*
- *Enforcement level "Strict": Every book and booklet has
  an imprint.*

A business policy is less formally structured and less concrete
as a business rule. For example:

"The books should be as budget-friendly as possible for the customers."

Figure 4.14 depicts the BMM concepts *Assessment* and *Influencer*.

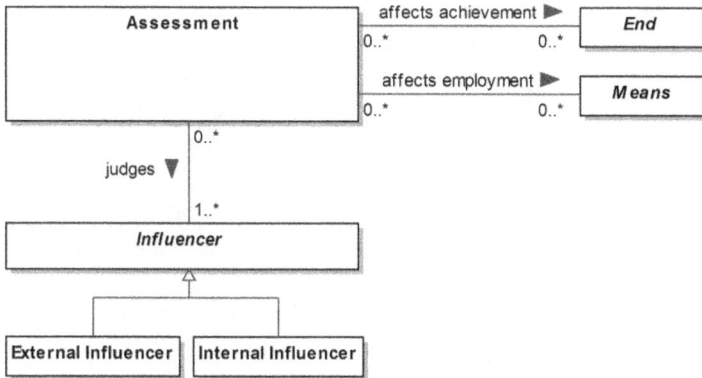

Figure 4.14: BMM Assessment & Influencer

An *Influencer* is someone or something that influences the business. Anything in the world influences something. For example, the famous stroke of a wing of a butterfly that could change the world.

In the business context, the BMM is a bit more concrete. It differentiates between internal and external influencers.

Internal Influencers are inside the business, and *External Influencers* are outside of the business. Influencers are concrete events like a systems engineering conference or the increasing demand for systems engineering in the manufacturing domain after the announcement of the Industry 4.0 initiative [Ger14].

Influencers could also be facts or best practices. In any case, the influencers are neutral. They are just there.

BMM defines some standard categories for internal and external influencers [BMM15]. The external influencer categories are:

Competitor	A rival enterprise in a struggle for advantage over the subject enterprise.
Customer	A role played by an individual or enterprise that has investigated, ordered, received, or paid for products or services from the subject enterprise.
Environment	The aggregate of surrounding conditions or Influencers affecting the existence or development of an enterprise.
Partner	An enterprise that shares risks and profit with the subject enterprise (or is associated with the subject enterprise to share risks and profit) because this is mutually beneficial.
Regulation	An order prescribed by an authority such as a government body or the management of an enterprise.
Supplier	A role played by an individual or enterprise that can furnish or provide products or services to the subject enterprise.
Technology	The role of technology, including its developments and limitations — there may be prerequisites for the use of technology; there may be enterprise activity that technology enables or restricts.

The internal influencers categories are:

Assumption	Something that is taken for granted or without proof.
Corporate Value	An ideal, custom, or institution that an enterprise promotes or agrees with (either positive or negative).
Habit	A customary practice or use.
Infrastructure	The basic underlying framework or features of a system.
Issue	A point in question or a matter that is in dispute as between contending partners.
Management Prerogative	A right or privilege exercised by virtue of ownership or position in an enterprise.
Resource	The resources available for carrying out the business of an enterprise, especially their quality.

The *Assessment* describes the impact of an influencer on the business. Figure 4.14 shows that an *Assessment* can judge one to many *Influences*, and an *Influence* is judged by zero to many *Assessments*. Without an assessment, an influence is a neutral element.

An *Assessment* affects the achievement of *Ends* and the employment of *Means*. It is the bridge between the *Influencer* and the *Ends* and *Means* of the business.

The BMM proposes the SWOT analysis as a standard approach for an assessment. However, you can use any other assessment method.

The SWOT analysis collects external and internal factors. Helpful external factors are opportunities (O), and helpful internal factors are strengths (S). Obstructive external factors are threats (T), and obstructive internal factors are weaknesses (W).

Figure 4.15 gives an overview of the SWOT Analysis method.

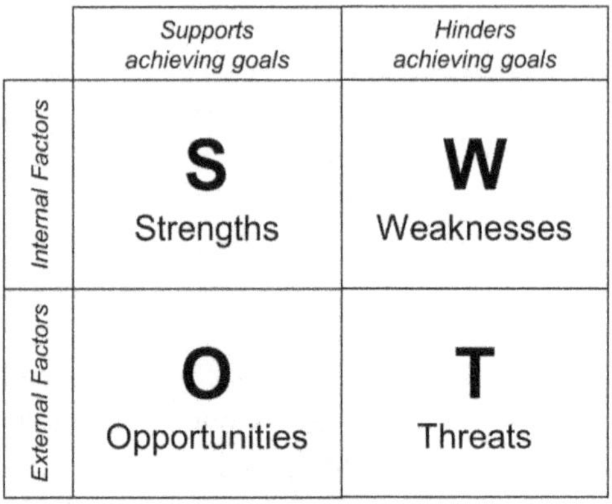

	Supports achieving goals	Hinders achieving goals
Internal Factors	**S** Strengths	**W** Weaknesses
External Factors	**O** Opportunities	**T** Threats

Figure 4.15: SWOT Analysis

The comprehensive list of (S)trengths, (W)eaknesses, (O)pportunities, and (T)hreats helps to spot how to use strengths to turn opportunities to good account, and how to improve weaknesses to avoid threats.

If you combine the internal and external factors, you get four fields of action that can be defined as new strategies for the business.

- Strengths & Opportunities: Pursue opportunities that fit to the strengths of the business.
- Strengths & Threats: Use strengths to prevents threats.
- Weaknesses & Threats: Deploy defense strategies to avoid the weaknesses.
- Weaknesses & Opportunities: Reduce weakness to utilize the opportunities.

Every business has strengths and weaknesses, and every market has opportunities and threats. Therefore you should analyze all four fields of action for your business.

The last pillar of the BMM is the *External Information.* They contain references to other business-related models for organization units, business rules, and business processes.

The business rule is also part of the *Directives* in the BMM. If defined outside in an external business rule model, for example, an SBVR model[2], the business rule directives in the BMM are the proxies of those business rules.

The organization units are the business structure that runs the business. An organization unit

- defines ends,
- established means,
- recognizes influencer,
- makes assessments,
- defines strategies, and
- is responsible for business processes.

The business processes are an essential part of the business, but out-of-scope of the BMM and therefore only part of the external information pillar. You can create business process models with the Business Process Model and Notation (BPMN) [BPMN13].

You can find more information about the BMM in the book *OCEB Certification Guide* [OCEB11]. BMM is part of the OMG-Certified Expert in Business Process Management program from the OMG (OCEB).

---

[2]Semantics Of Business Vocabulary And Rules [SBVR17]

The BMM specification is available for free on the OMG
website: www.omg.org/spec/BMM.

Good moves on the engineering playfield

Clarify your vision and mission with the business motivation
model. If you already know it well, it should be easy and is
a validation of your good understanding of your business. If
not, you recognize your gaps and can use the BMM to close
them.

4.6 Base Architecture

The base architecture covers architectural, technical and organisational decisions that are preset for a product or service development project [We15].

Nearly no development project starts from scratch. A car has four wheels, an aircraft two wings, and a smartphone a touch display. These are technical decisions that are typically not part of the design space but are already fixed at the beginning of the project as part of the requirements.

The base architecture sets this starting point and opens or narrows the design space for the system or service under development.

Figure 4.16 shows the sketch of a base architecture of a forest fire detection system (FFDS) taken from the book [We15].

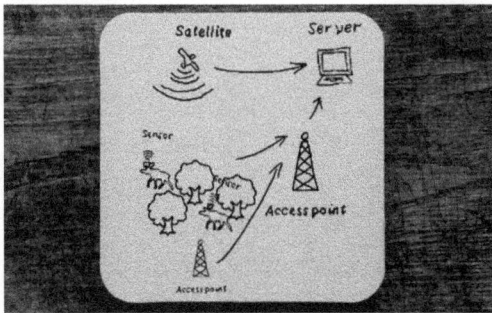

Figure 4.16: Sketch of the Base Architecture of a Forest Fire Detection System

The beer mat architecture is a simple form of the base architecture. You can also create a comprehensive architecture documentation. It depends on your project-specific demands what is appropriate. The format is independent of the fact that the base architecture exists and is always there.

The FFDS base architecture in figure 4.16 presets that the system has to use satellite observation and sensors attached to animals in the forest (the movement patterns of the animals indicate forest fires). The sensors are wind together by access points to a sensor network and connected with a central server.

These architectural and technical decisions are part of the requirements of the engineering project. They constrain the design space for a solution.

The base architecture is often a slight abstraction of the predecessor product on the market. The next product generation is similar to the previous one with a few improvements, and it is seldom wholly different.

Having a description of the base architecture is important. First, it is an input for your product development. The requirements depend on the base architecture. For example, the FFDS system has requirements about the satellite communication or the animal sensors. Without the base architecture, an animal sensor would be part of the solution and not a requirement.

Second, the base architecture is also an excellent source to look for significant and disruptive innovations. Changing the base architecture means to change the architecture of your product radically.

And third, the base architecture is always there. So, it is a good idea to make this implicit knowledge explicit to avoid misunderstandings that lead to ineffective communication and cost-intensive errors.

Another illustrative example for a base architecture is a system to play recorded music. In 1887 Emil Berliner invented the gramophone (figure 4.17) [Pa87].

Figure 4.17: Emil Berliner with his gramophone [ImgBe]

Figure 4.18 depicts a simple sketch of the base architecture of the gramophone.

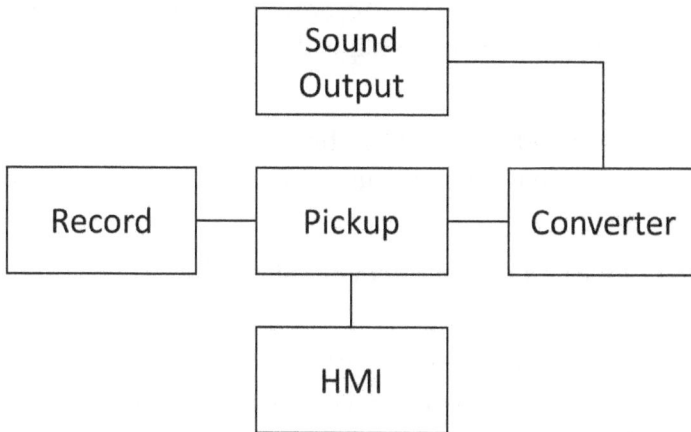

Figure 4.18: Base architecture of a record player

On that level of abstraction, the base architecture is valid until today. Although, of course, several improvements of the concrete products were made. Nowadays, a gramophone respectively records player, for example, does not have a crank and zinc-based records anymore.

However, in the mid 80ies, a new technology came up: the CD player, and a few years later CDs and not records were the mainstream technology for music. The CD was the first strike against the base architecture. Although, on a high level of abstraction it is similar to a record.

At the latest, MP3 and the streaming technology deflated the base architecture. It was a disruptive innovation that changed not only the technology but also the business models. You rent access to music libraries on a monthly basis instead of buying the music.

Typically, a change of the base architecture has a significant impact on the organization and business. It can change supplier chains, business processes, market segments, business models, and many more. According to Conway's law, a communication structure is aligned with the base architecture (see section 3.3).

The Zigzag pattern (see next section 4.7) clarifies the relationship between base architecture, requirements, and product architecture.

Good moves on the engineering playfield

- Make your base architecture explicit.
- Regularly reconsider the base architecture if it has some innovation blocker.
- Align the base architecture with the communication respectively organization structure of your organization.

4.7 Zigzag Pattern

The zigzag pattern describes the relationship between the problem space and the solution space, or in other words between requirements and the product architecture.

Requirements should not anticipate the solution. They describe the *What*, and the product architecture describes the *How*. Sounds easy, but it is not easy.

Requirements are solution-free, and they contain solution aspects at the same time. It depends on the viewpoint and level of abstraction.

Let us assume that you have solution-free requirements (I argue that those requirements are not viable in practice). Now you derive a product architecture that satisfies the requirements and you get the typical what/how-pair. For example, you have requirements for a transportation system for people, and you derive a product architecture that specifies a car. The solution *car* leads to new requirements that contain aspects of that solution, for example, requirements about the engine of the car and self-driving capabilities. You have not those requirements if your product architecture specifies a bike.

These requirements are on another abstraction level and solution-free from the viewpoint of that level, but they contain solution aspects from the previous level.

Again, you derive a solution from the requirements, for example, a hybrid engine, and again, that solution leads to new requirements, and so on. All in all the logical steps represent a zigzag pattern (figure 4.19).

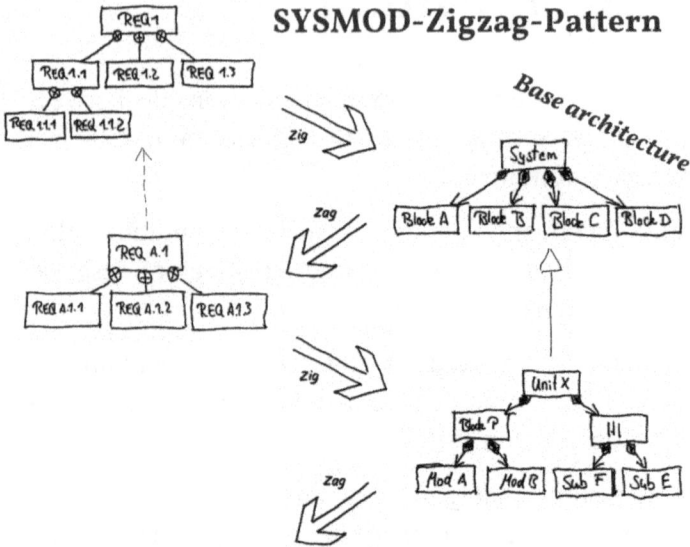

Figure 4.19: SYSMOD Zigzag Pattern [[We16a](#BibWe16a)]

In practice, requirements always contain some solution aspects. Unfortunately, they are often implicit and are one of the causes why requirements are a sore spot of many projects. The product architecture above your top-level requirements is called the base architecture of your product (see section 4.6).

Good Moves on the Engineering Gameboard

The base architecture is always there. Make it explicit. Even a one-pager sketch of the architecture is already valuable. The best level of detail depends on the specific demands of the project.

4.8 Design Thinking

Design Thinking is an approach to solve problems creatively. The origin of Design Thinking lies in the 1960s [Ar65]. An important recent publication about Design Thinking is *Change by Design: How Design Thinking Transforms Organizations and Inspires Innovation* from Tim Brown [Br09].

In the business and engineering domain, it is well-known and applied for around 15 years. The main drivers were the d.school at the Stanford University, the company IDEO and the Hasso Plattner Institute in Potsdam, Germany.

Design Thinking does not only elaborate a solution, but it also has a strong focus on the problem space. It allows the redefinition of the initial problem statement to find the real challenge that satisfies the stakeholder needs. Design Thinking does not ask what the users want, but what the users need. An illustrative example for finding the real problem is the quote by Henry Ford: *"If I had asked my customers what they wanted they would have said a faster horse."*

The quote hits the point. Therefore I also use it in section 4.9 about complex product engineering where we also cover the problem space.

The problem statement is co-developed with the solution in an iterative process. A first problem statement leads to first solution attempts that again clarify the problem statement, and so on.

The user is always in the center of the Design Thinking process. Besides the visible problem, Design Thinking also addresses the invisible emotional needs of the user.

It combines problem analysis and solution elaboration, and synthesis, divergence, and convergence. Starting from a problem you open the space for user needs to finally pick the most relevant one and give a clear statement of the challenge. Starting from the challenge, you reopen the room for as many solutions as you can, finally selecting the best one.

Figure 4.20 gives an overview of the Design Thinking approach.

Design Thinking Process

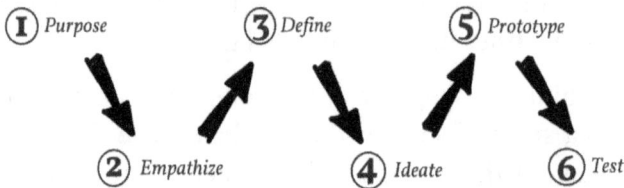

Figure 4.20: Design Thinking Process

Each step has potential jumps back to one of the previous steps. It would blast the figure to show all of them. So, you only see the forward direction.

The Design Thinking process described here has six steps. There are other slightly different versions of the Design Thinking process with more or fewer steps.

Each step comes with a set of methods to achieve the goals of the step.

Purpose: First you must agree on the purpose of the Design Thinking undertaking. It is the initial statement of the problem that you would like to solve. It is the starting point to set the context for the next *Empathize* step. The *Define* step sets the final problem statement.

Empathize: Work out the needs of the users with a particular focus on the emotional level. Reframe your understanding of the problem in the context of the users. In this step, you typically do user research. It is an analysis task that leads to many data to be processed in the following steps.

Define: Give a clear statement of the challenge that should be addressed by the solutions developed in the next steps by synthesizing the output of the *Empathize* step.

Ideate: Brainstorm a broad range of ideas for solutions to the defined challenge. Every unusual idea is welcomed. Filter out the most promising solution ideas.

Prototype: Create prototypes of your ideas to interact with them and to learn from them. Prototypes can be very simple like drawings for user interfaces instead of programming or building them. Relevant is the physical experience in the user environment of the idea and not to overcome technical challenges, or to build a real looking system.

Test: Test your prototypes and use the findings to refine the outcomes of the previous steps of the Design Thinking process. If possible, you test your prototypes together with the real users.

Each step has a set of methods like personas, empathy maps, SCAMPER, or DeBono. Check one of the Design Thinking books about detailed descriptions of these and many other methods. In the following we have a brief look at the SCAMPER method.

SCAMPER is an acronym for *Substitute, Combine, Adapt, Modify, Put to another use, Eliminate,* and *Reverse.* It is a thinking model to find unusual solutions. Each letter stands for a technique to modify your product or product idea that potentially can improve it.

- **Substitute**: Replace a part of your product with another part.
- **Combine**: Combine existing products or services to a new one.
- **Adapt**: Use your solution idea to solve another problem.
- **Modify**: Modify one part or aspect of your product.
- **Put to another use**: It is similar to *Adapt*, but the usage scenario is entirely different.
- **Eliminate**: Remove a part of the product or service. It is a lean concept to remove waste, i.e., things that do not add any value for the customer (section 4.13).
- **Reverse**: Think your solution the other way round, i.e., completely against its original purpose. The new perspective can lead to new exciting insights.

The Design Thinking process perfectly fits into an engineering development process (see section 4.9). The outcome of the process is a proven idea of a product or service. It includes the system idea and some requirements as inputs for the engineering process to finally develop the product.

Good moves on the engineering playfield

- Perform a design thinking workshop to clarify the problem you want to solve with your products.
- Conduct the design thinking workshop with people from many different fields.

4.9 Complex Product Engineering

This section presents a general-purpose methodology for the development of complex products. For sure, it is not possible to give a complete and detailed step by step development procedure that is valid and effective for all domains and organizations. The given methodology describes only a toolset of essential steps on a high level of abstraction.

You should not miss these essential steps in an engineering project. That is why they are called *essential*. You have high degrees of freedom in which detail and format you perform them, and of course, you must add more steps specific to your domain and engineering style.

Figure 4.21 depicts the outcomes of the steps as a concept model (see also section 4.5 about concept models).

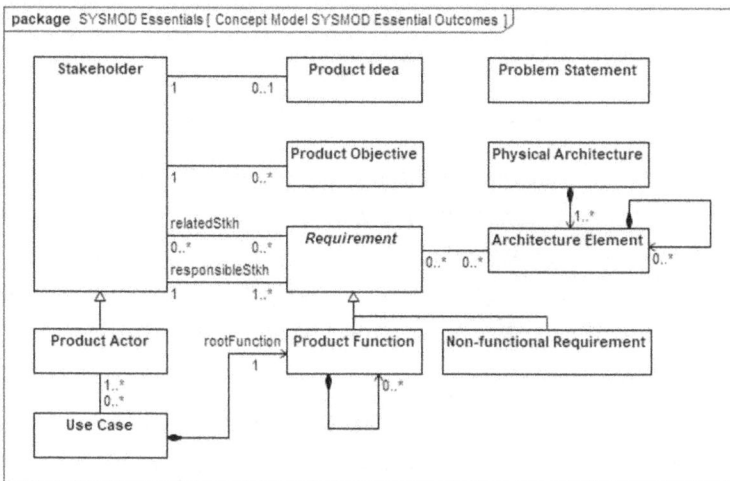

Figure 4.21: SYSMOD Outcomes

The methodology is derived from the SYSMOD methodology [We16a]. SYSMOD stands for Systems Modeling and is a methodology in the context of MBSE (see section 4.11). The derived methodology is called *SYSMOD Essentials* and is independent of modeling. It does not describe how to create a model, but on an abstract level which outcomes are essential, how they are connected, and how to work them out.

Figure 4.22 gives an overview of the essential steps. The following section explains them in detail.

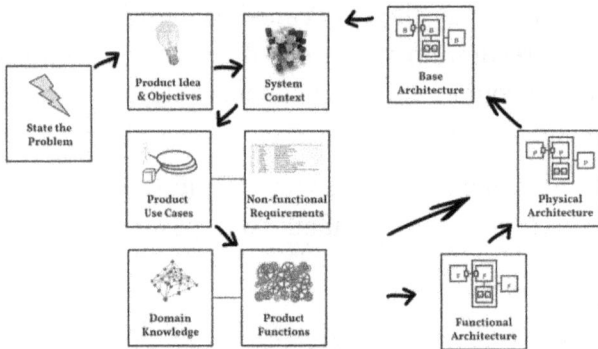

Figure 4.22: Essential SYSMOD steps

You may have noticed that I use the term *Product* as well as *System*. In my terminology, a product is a special kind of a system that is sellable and has a particular focus on the market. However, in most cases, the distinction is not relevant, and I use the terms often as synonyms.

In the following, I describe the steps of SYSMOD Essentials in a logical order. Nevertheless, it is not a process, and the steps could be executed separately or in a different order. Like a puzzle: you can work everywhere you want, but in the end, you get a complete and consistent picture. Typically, you often revisit a step and create the outcome incrementally.

4.9.1 State the Problem

4.9.1.1 Purpose

Move one step back before you start spending effort on the product development, and analyze the problem that the product shall solve.

4.9.1.2 Main Description

Most engineers have the "rush to solution" syndrome. They think in solutions and love it to work them out. However, what about the problem behind the solution? The problem is the starting point and should be carefully stated. Everything else depends on it and must change if the problem changes, or is useless if the problem statement is wrong.

It is good to look twice on it, and to be sure about the answer to the question "Which problem do we solve and it is the right problem?". Henry Ford said, *"If I had asked my customers what they wanted they would have said a faster horse."*[3]. The quote implies two points: First it confirms that it is important to state the right problem, and second that even the customer does not know the real problem statement.

The Design Thinking methodology includes elements to elaborate the problem statement. Section 4.8 covers Design thinking in more detail. Besides the problem statement, Design Thinking also elaborates prototypes, the product idea, and objectives. It is a good predecessor of the development project.

---

[3]By the way it is not proven that Henry Ford had said this. However, it is a great sentence.

4.9.2 Describe the idea and objectives of the product

4.9.2.1 Purpose

All participating parties of the development project should know the idea and the objectives of the product to ensure that the right decisions are taken along the way toward the final product.

It is not to be taken for granted that the project members well know the product idea and objectives. They must be communicated actively.

If you know the objectives, you find a solution even if some distances of your way are unknown, unsafe, or full of obstacles. If you do not know the objectives, but all the rules of a perfect development process, you reach a destination that probably does not match the real needs.

4.9.2.2 Main Description

There are various sources for the product idea and objectives. Be it a genius thought, a given contract, an output of product management work, or anything else.

Typically, this step works up the somewhere else in the organization stated notion and objectives of the system to artifacts for the communication inside the system development project. Not only the product, or portfolio, or whatever management should know it, but especially the engineering team. You can do that with a workshop and invite all or most of the relevant people.

A tool for the workshop can be the product box [We16a]: You start with an empty box that represents the package of your product. Together with all workshop participants, you put the captions, figures, and the text on the box that covers, beside others, the idea and objectives of the product that is virtually inside the box.

Another tool is the product vision board by Roman Pichler [Pi16]. Figure 4.23 depicts the *Product Vision Board* template. Roman Pichler calls it a board. Actually, it is another canvas (see section 4.3).

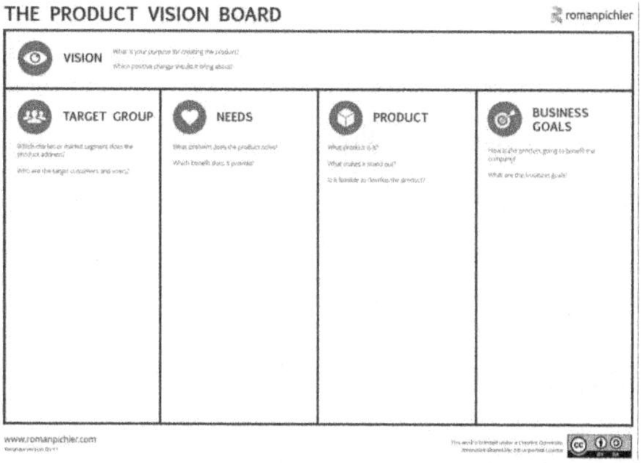

Figure 4.23: Product Vision Board (Source: [Pi16])

The sections of the board cover the product idea and objectives. Additionally, it also covers aspects like stakeholders (see next section 4.7.3).

Figure 4.24 depicts a card to document the product idea in a workshop context. The card is part of the SYSMOD Toolbox Card Set [We18]. The cards provide templates for the essential information, can easily be used in workshops, and enable a structured transition path for the workshop outcome into a model or document. Other essential steps of the SYSMOD Essentials methodology also contain some cards from the card set (see other step description sections).

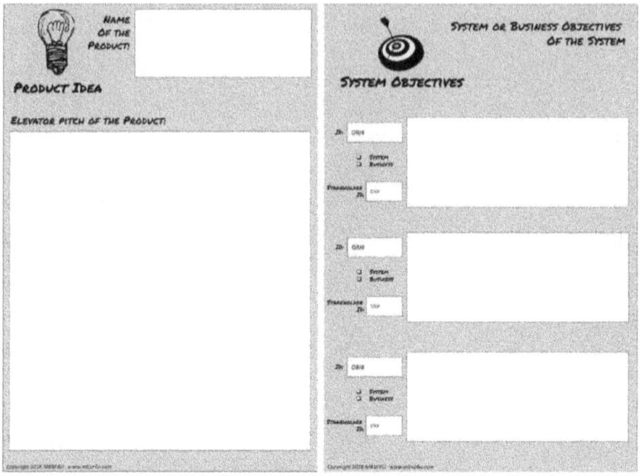

Figure 4.24: Workshop cards "Product Idea" and "Product Objectives"

Ensure that the product idea and the objectives are actively communicated in the team and to the stakeholders, and repeat it whenever you think it needs to be refreshed.

4.9.3 Identify Stakeholders

Identify all individuals and organizations that may have requirements or an interest in the product.

4.9.3.1 Purpose

It is decisive for the success of the project that you sufficiently consider the concerns of all stakeholders. A forgotten stakeholder can break the project.

4.9.3.2 Main Description

The list of stakeholders is initially elaborated in a workshop and continually reworked during the project. More requirements analysis and product architecting detects and leads to more stakeholders.

The stakeholder's concerns are a source for formal require-ments that must be satisfied, for example, based on a contract. The stakeholders are also a source for informal needs that are not explicitly stated but also important to the product or development team.

We document the name, the concerns, and a general descrip-tion of the stakeholder. In order to be able to make contact, we also store a contact information (email, availability, and so forth). To prioritize the typically very long list of stakeholders, we classify the priority and the effort to consider the stake-holder. That enables a 2-dimensional prioritization to cluster the stakeholder in groups (figure 4.25).

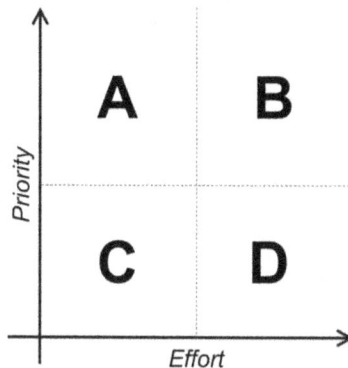

Figure 4.25: 2-dimensional stakeholder priorization

You manage the stakeholders based on their group assign-ment:

- A - Mandatory stakeholders who typically cost much effort to get their requirements and needs. Consider the time in your project plan.
- B - Mandatory stakeholders who are typically easy to handle.

- C - Not very important stakeholders who do not cost much effort. Consider them with a medium priority.
- D - Not very important stakeholders who cost much effort. Re-evaluate the group assignments to be sure, and do not consult the stakeholders if they are still in group D. Only consider them if you are bored and have enough time.

Figure 4.26 shows a card to record stakeholder information in a workshop scenario [We18].

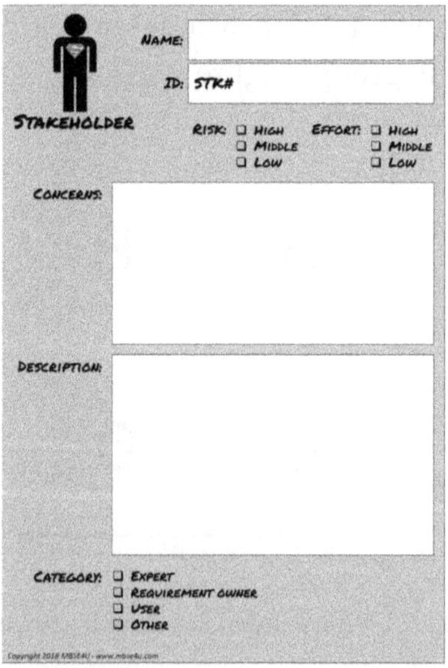

Figure 4.26: Workshop card "Stakeholder"

4.9.4 Define requirements

4.9.5 Purpose

The requirements are the contract between the engineering project and the stakeholders of the product.

4.9.6 Main Description

The requirements specify the features of the product. Depending on the abstraction level, the requirements are high-level user requirements, or system requirements, or component requirements (see section 4.7).

The stakeholders are the source of the requirements. You receive the requirements as documents, as an outcome of workshops, a result of a discussion, in the form of a presentation, and so on. The elicitation and management of requirements is a broad topic on its own and out of the scope of this book. You find many books about requirements engineering for your preferred engineering style at your book dealer.

Figure 4.27 shows a card to document a requirement in a workshop scenario [We18]. The card covers basic information about a requirement:

- a name, for example, *Service reliability*
- an unique identifier
- a text that defines the requirement. It can also be a reference to a model element (see section 4.12).
- a category: functional, non-functional, constraint, or other
- a reference to the responsible owning stakeholder of the requirement

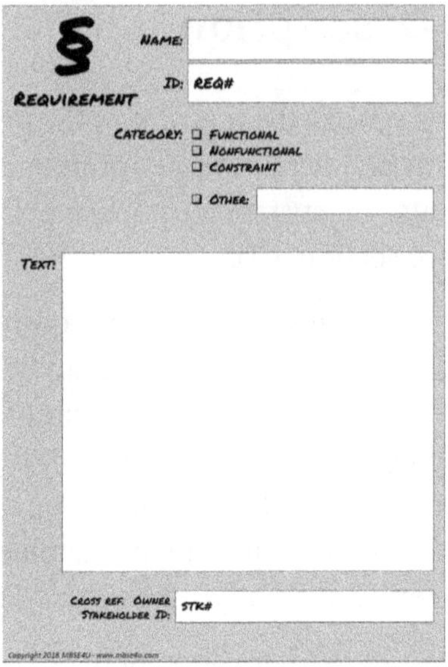

Figure 4.27: Workshop card "Requirement"

4.9.7 Describe the Product Context

Identify the users and other external entities that interact with the product.

4.9.7.1 Purpose

The product context defines the environment of the product that needs to be considered, defines the product boundary, and depicts the required interfaces between the product and its environment.

4.9.7.2 Main Description

The product context depicts all elements from the environment of the product that interacts with the product. These elements are called product actors. The obvious ones are the users of the product and external systems with explicit interfaces. Less obvious product actors, but equally important are, for example, environmental effects like temperature or mechanical systems like a floor space, or a wall. Other product actors are external entities that are only connected with the product to exchange electricity, fluids, or other infrastructure things.

In addition to the list of product actors, the product context describes interfaces and relevant flows of items between the product actors and the product.

The human actors and the humans or organizations behind the non-human actors are also stakeholders of the product (see section 4.7.3).

Figure 4.28 depicts a card from the workshop card set to document a single actor in a workshop scenario [We18]. The cross-reference section can link the actor to appropriate stakeholders.

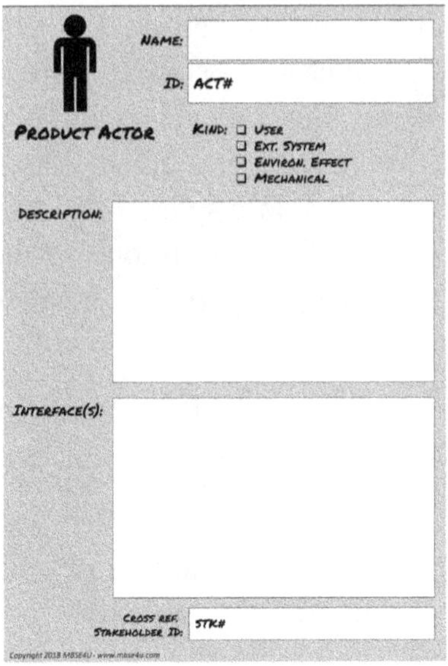

Figure 4.28: Workshop card "Product Actor"

4.9.8 Define Use Cases

Define all services provided by the product for the actors and stakeholders of the product.

4.9.8.1 Purpose

The use cases are the services and the essential ends of a product. The product is developed and operated to achieve these services.

4.9.8.2 Main Description

The use cases provide an outside-in view on the product functions from the perspective of the product actors. That enables the engineering team to build a product that really satisfies the needs of the product actors.

An inside-out perspective of the functional requirements also considers all functions, but with the risk that they do not fit to the usability needs of the users.

A good example in most cases is remote controls for a projector (figure 4.29). They provide all required functions, but not from the perspective of the users, but more from the perspective of the engineers. The remote control provides lots of buttons, and the primary use cases (on/off, mute, freeze) are hard to find.

Figure 4.29: Remote Control of a projector

The use case description should at least include

- the associated product actors,
- the trigger that starts the use case,
- the result of the use case,
- a brief textual description (2-5 sentences),
- the non-functional requirements that are relevant to the use case.

Figure 4.30 shows a card to document a use case in a workshop scenario [We18].

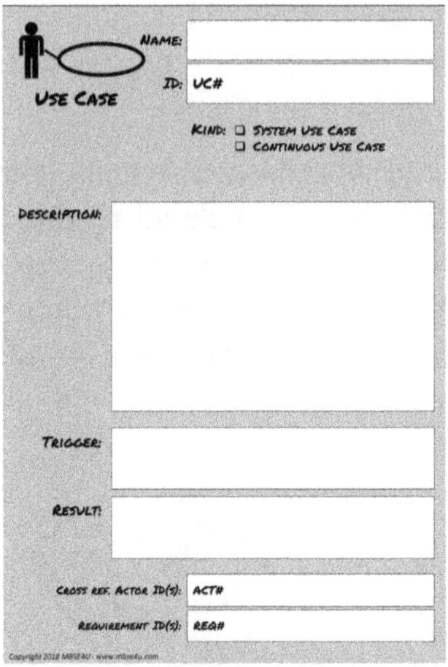

Figure 4.30: Workshop card "Use Case"

4.9.9 Describe Product Functions

Describe the behavior of the product by a set of functions.

4.9.9.1 Purpose

The product functions are the functional decomposition of the use cases including the order of execution and the object flow between the product functions.

4.9.9.2 Main Description

A use case specifies beside others a name, a trigger, and a result. The product functions specify the behavior of a use case. The specification could be a rough description or an unambiguous and detailed specification of the functions. The level of detail depends on the needs of your project.

Each action within a product function is again specified with another function. The product functions that need no further refinements have no included actions.

A product function description includes the input and output objects of the function. The relationship of an output object of a function to an input of another function is called object flow.

It is a good practice to separate the functions that are responsible for the input and output of objects from and to the product actors from all the other functions. Those input/output functions depend on the interface technology that is typically more unstable than the core functions and are less dependent on the specific domain.

Figure 4.31 depicts a card to document a product function in a workshop scenario [We18]. The card does not cover the execution order of the actions within a product function. The execution order can be depicted in a workshop by putting the cards in an order on a pin board. However, depending on the level of detail you probably do not need the execution order information at this stage of your engineering process.

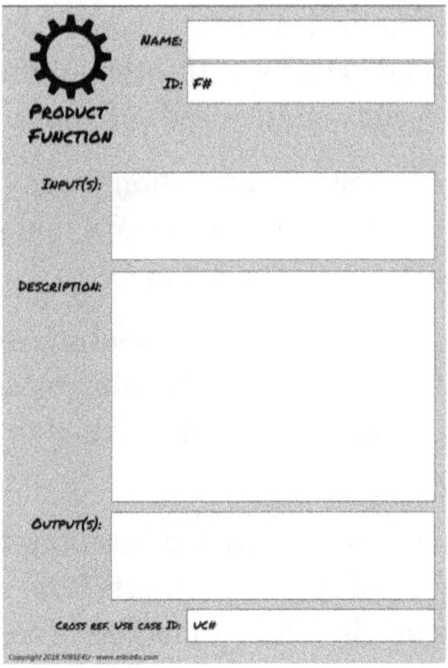

Figure 4.31: Workshop card "Product Function"

4.9.10 Model the Domain Knowledge

Define the terms of the domain from the perspective of the product with a concept model.

4.9.11 Purpose

The domain knowledge defines the semantic and structure of the domain objects that are used by the product.

4.9.12 Main Description

The product knows objects of the domain. Imagine your system is a person and you ask it about domain objects.

For example, you can ask a forest fire detection system (FFDS):

You: "Do you know the concept of an operator?"

FFDS: "Yes. An operator is one of my users and has an ID, a name and list of active tasks."

You: "Do know the concept of fire?"

FFDS: "Yes. A fire has a severity, position, and size."

The domain knowledge defines the knowledge of the product about the domain.

You can derive the domain objects from the object flow of the product functions. If an object is the input or output of a function, the product knows the concept of that object.

Especially, if you separate the input/output functions from the core functions, you get two kinds of domain objects:

1. The context objects are entities that are exchanged between the product and the product actors.
2. The product objects are product objects that are used only inside the product.

The domain knowledge is also known as *Concept Model* of the product (see section 4.5).

Figure 4.32 shows a card to document a single concept [We18].

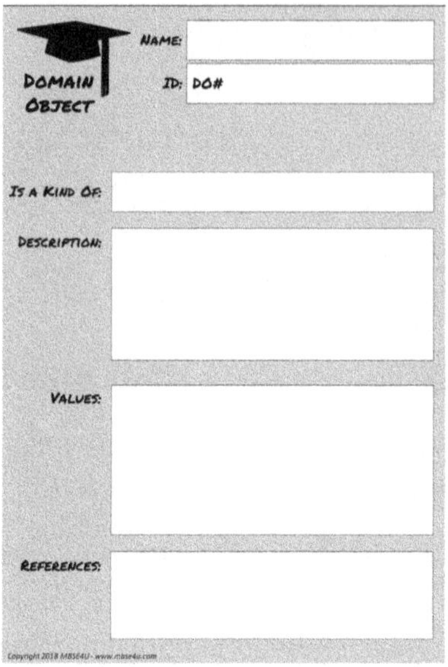

Figure 4.32: Workshop card "Product Function"

4.9.13 Specify the Product Architecture

Specify the physical architecture of the product.

4.9.14 Purpose

The product architecture is the specification of the architecture of the product on the system level.

4.9.15 Main Description

Every engineering discipline like mechanical, electrical, or software, has its own methods and tools how to specify and document the work. The product architecture is one abstraction level above the concrete engineering disciplines, and specifies the overall architecture of the product.

The product architecture documents the fundamental decisions about the structure and technologies of the product made by the product architects. It is the leading architecture that provides requirements and the context for the concrete engineering disciplines.

Figure 4.33 depicts a card to document a single part type (=block) of the product architecture in a workshop scenario [We18].

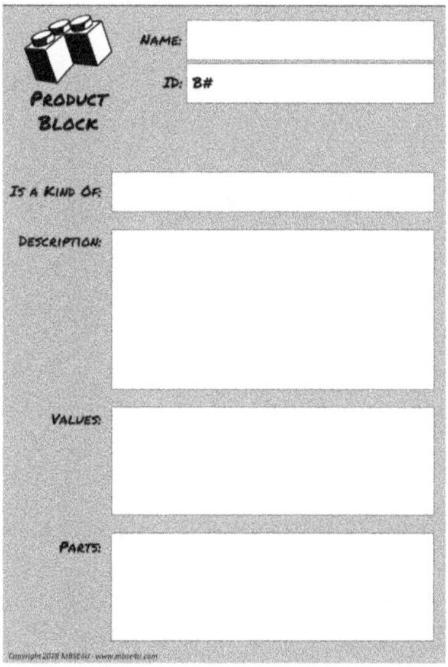

Figure 4.33: Workshop card "Product Block"

It does not show the relationship between the parts. In a workshop you can pin the cards on a board an draw the relationships between them. Later you transfer the information into the final product architecture documentation tool that typically stores much more details.

I recommend a SysML model to specify the product architecture, and also all the other artifacts covered in this subchapter.

Good moves on the engineering playfield

- Apply a product engineering methodology that explicitly creates and uses the essential outcomes.
- Use models to manage the requirements and architecture engineering information.
- Clarify the terms of your methodology. It is more important that everyone in the project has the same understanding of the concepts than copying a predefined methodology.

4.10 Functional Architecture

A functional architecture is an architecture independent of the technical implementation of the system, more stable across product families and generations and guidance to derive a sustainable physical architecture.

The functional architecture is defined as an "Architecture based on functional elements, functional interfaces, and architecture decisions." [We15].

Technology changes, but the domain-related functions are stable. For example, the high-level functions to play music are the same for a record player as for a music streaming service. None of the leading music streaming providers are one of the leading record player companies from the past. First, you should focus on providing excellent services to your customers, and secondly, you can focus on the technologies that implement the functions.

At first glance, it seems trivial to describe functions independently of a technology. On the second view, you realize that it is not that simple. What exactly does it mean to be independent of technology?

The reference point is the base architecture (see section 4.6). The functions of the functional architecture may depend on the technologies specified in the base architecture. You achieve the technology-independency of the functional architecture by deriving the architecture from the requirements assuming that they are correctly specified regarding the independence of the solution space.

The Functional Architecture for Systems Method (FAS Method) [We15] provides an intuitive way to derive a functional architecture from a use case analysis (see section 4.7.6).

The system functions of the use case descriptions are grouped based on user-oriented usages of the system. This structure fits in perfectly with the requirements engineering tasks to develop the user and system requirements. However, the product architect needs an additional view on the system functions that groups the functions by common character-istics (cohesion principle). Functions that do similar things should be in the same group. That is - simply said - the basic notion of the functional architecture.

Figure 4.33 depicts the relationships between physical archi-tecture, functional architecture, and the use case functions.

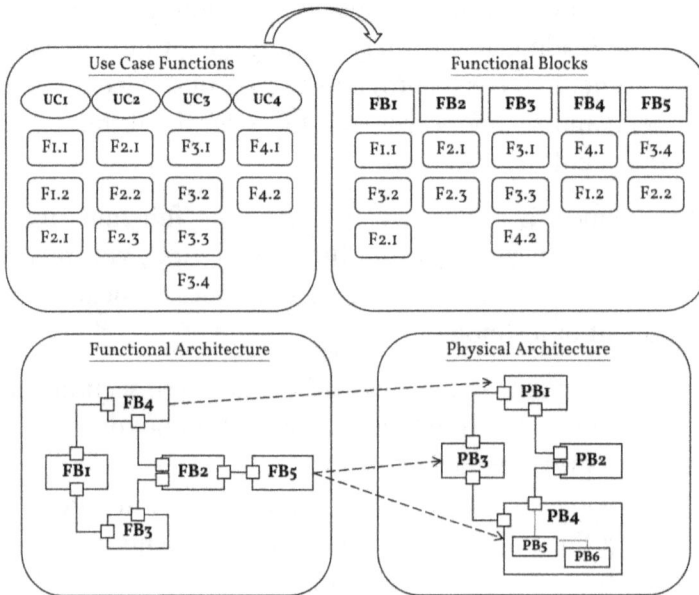

Figure 4.33: Functional Grouping

The use case functions are functional requirements grouped by the use case criteria (upper left in figure 4.33). A functional block is an element of the functional architecture and groups a set of use case functions following the criterion of cohesion (upper right in figure 4.33). A functional block is allocated to a physical block of the physical architecture that specifies the implementation of the functions and other features of the physical element (lower part in figure 4.33).

Figure 4.34 provides an overview of the FAS method.

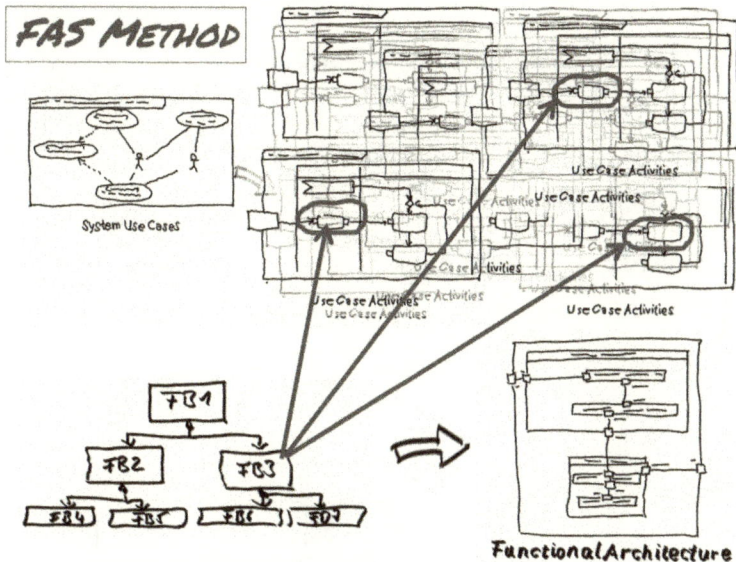

Figure 4.34: Overview FAS Method

The upper part of the figure shows use cases and flow diagrams of the decomposed use case functions (SysML activity diagrams). Step 1 of the FAS method is the grouping of the use case functions. Functions are in the same group if they do similar things (cohesion principle). The grouping results in a set of functional blocks (lower left part in figure 4.34).

They are the building blocks of the functional architecture. They are connected in the functional architecture if functions of one block provide an output that is the input of functions of the other block[4] (lower right part in figure 4.34). You find a detailed description of the FAS method in the book *Model-Based System Architecture* [We15].

Figure 4.35 depicts an extract of the functional architecture of the forest fire detection systems.

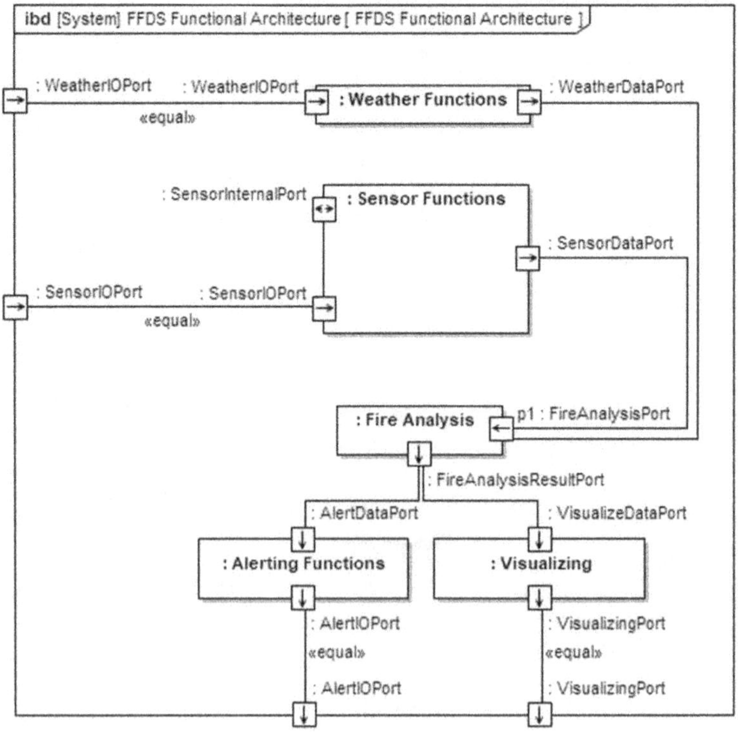

Figure 4.35: Overview FAS Method

[4]More precisely the functional architecture does not connect the blocks, but functional properties. However, these are details of the language SysML and out-of-scope of this book.

The frame represents the border of the system. The boxes on the frame are functional interfaces to external entities. The rectangles within the frame are typed by functional blocks and represent each a set of coherent functions. Again, the small boxes on the rectangle borders represent functional interfaces.

The functional interface *SensorIOPort* at the lower left side of the frame specifies the input of things detected by sensors of the FFDS. The things flow into the functional part called *SensorFunctions* where functions process the input and transform the physical things to data. The data flows from the *SensorFunctions* to the functional part *FireAnalysis*. Their functions analyze the data and send their output to the *VisualizingFunctions*, and, if necessary, to "AlertingFunctions".

All the functional parts are independent of the technical implementation except for the technologies that are preset in the base architecture (section 4.6).

You derive the physical product architecture from the functional architecture. As a rule of thumb, a functional part shall ideally be implemented by only one physical part of the product. Since the functional parts are created based on the cohesion principle, it is very likely that a change request only affects a single functional part and a therefore also only a single physical part. It is a minimal impact on the product.

Of course, in practice, you are not able to do a strict 1:1 mapping of functional parts to physical parts, and other requirements like safety can contradict this principle. However, you should aim for it, and all relevant deviations point to areas in your architecture that need special attention.

Figure 4.36 shows an allocation matrix that specifies the mapping of top-level functional blocks to physical blocks of the FFDS.

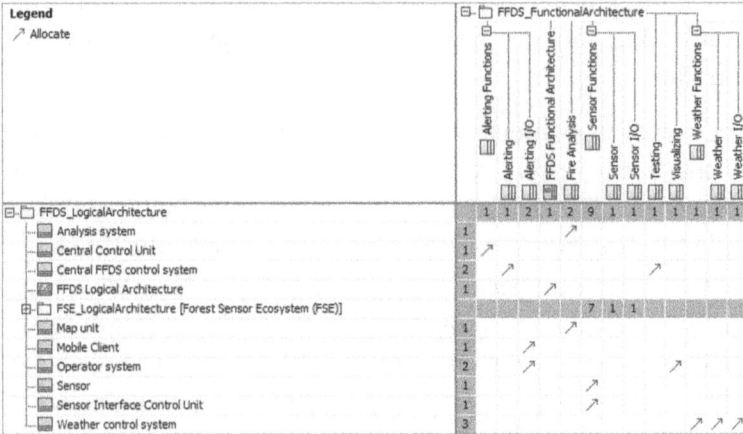

Legend: ↗ Allocate		Alerting Functions	Alerting I/O	FFDS Functional Architecture	Fire Analysis	Sensor Functions	Sensor	Sensor I/O	Testing	Visualising	Weather Functions	Weather	Weather I/O
FFDS_LogicalArchitecture		1	1	2	1	2	9	1	1	1	1	1	1
Analysis system	1				↗								
Central Control Unit	1	↗											
Central FFDS control system	2		↗								↗		
FFDS Logical Architecture	1			↗									
FSE_LogicalArchitecture [Forest Sensor Ecosystem (FSE)]							7	1	1				
Map unit	1			↗									
Mobile Client	1		↗										
Operator system	2		↗								↗		
Sensor	1					↗							
Sensor Interface Control Unit	1					↗							
Weather control system	3										↗	↗	↗

Figure 4.36: Functional to Physical Mapping

For example, the functional block *Fire Analysis* is allocated to the physical block *Analysis system*. The *Analysis system* is a software-centric system that implements the functions included in the functional block *Fire Analysis*.

Figure 4.37 depicts an extract of the physical architecture of the FFDS. It covers mainly technical concepts and principles, and not concrete, detailed technical specifications. Such a physical architecture is also called a logical architecture. The *Analysis system* is connected with the central control unit and with external research systems.

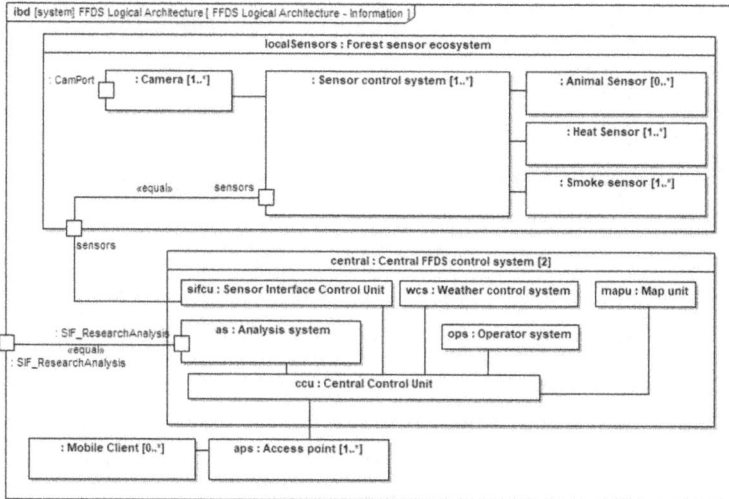

Figure 4.37: Physical Architecture of the FFDS (extract)

Good moves on the engineering playfield

- Value functions more than the technology. They are the essential part of your product.
- Create a functional architecture to derive a physical product architecture, or
- Use a functional architecture to assess an existing physical product architecture.

4.11 Model-Based Engineering (MBE)

The discipline Model-Based Systems Engineering (MBSE) - MBE with an "s" - is getting more and more popular. The primary language of MBSE is the OMG Systems Modeling Language (SysML) [SysML17]. SysML is only one of many possible modeling languages in the context of MBSE.

The intense use of models is only one piece of MBSE. The others one is systems engineering itself. Systems Engineering is a wide-spanning discipline that covers many aspects of product engineering. Chapter 3.5 gives an introduction to systems engineering.

Model-based engineering (MBE) - without "system" (s) - goes a step further. The paper [PLM4MBSE15] defines MBE as a combination of lifecycle spanning management of product data (Product Lifecycle Management (PLM)) and the formal, model-based description of systems (MBSE).

You need a consistent, holistic, and connected model of your product as a foundation to cope with the challenges of complexity and dynamics that face many companies nowadays.

Models are crucial to the engineering of complex products. Without models, you cannot effectively manage all the dependencies between your engineering artifacts. But what is a model? Seems to be an easy question, but actually it is not! Which criteria classifies, for example, a SysML model as a model and an office document like a spreadsheet or text document not as a model?

If you search for definitions of the term *Model*, you find statements like "A model is an abstraction of something.".

Often the "something" is replaced by "real thing". To be precise, it should also include the product under development, i.e., the thing that does not exist yet.

The "something" can also be a problem or any other thing. In any case, that definition does not exclude text documents. Text can also be an abstraction of a thing. Just write a descriptive text about it.

Another typical feature of a model is the separation of repository and views. The repository is the storage of the model data. The view is the representation of the data. If separated you can provide specific views for different stakeholders. Typically, SysML model elements are often depicted in several diagrams, i.e., in different views. Figure 4.38 depicts a simple context diagram and a use case diagram of a forest fire detection system. The actor *Operator* is shown in both diagrams, but there is only one single entity of the actor in the repository.

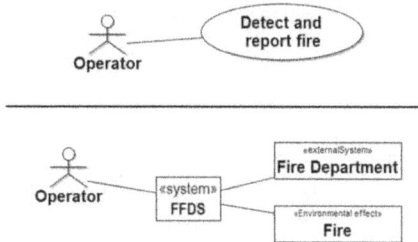

Figure 4.38: Two different views, one single entity *Operator*

However, you find the same separation of views and repository in office documents. For example, think of all the different diagram types to show worksheet data. The separation of repository and view is not a clear discriminator between office documents and SysML models.

The criterion that classifies a model of a specific domain is "The semantic and abstract syntax of the modeling language covers the concepts of the domain.". Here, the domain is MBE. The semantic and the abstract syntax is the vocabulary of the modeling language. For example, requirements and architecture blocks are part of the SysML language. However, you can not find these concepts in office document languages.

Office document languages cover concepts like the headline, paragraph, or bold font. An office document language is a modeling language for office document, and an office document is a model in that context, but not an MBE model.

SysML is an MBE modeling language, but also proprietary languages like included in engineering tools IBM DOORS, Modelica, Simulink, and others are languages of the MBE domain.

Only if the modeling language covers concepts of the domain, it enables useful analyses and visualization of the model.

Back to MBSE - what makes MBE important for the new engineering game?

The paper *10 theses about MBSE and PLM - Challenges and Benefits of Model-Based Engineering (MBE)* [PLM4MBSE15] postulates the following theses about MBE:

1. MBE is the enabler for the "Internet of Things" and "Industrie 4.0".
2. Product liability and functional safety regulations are a driving factor for MBE.
3. Future PLM systems need a holistic view of a product as a multidisciplinary system.
4. Early design decisions must be logically, and functionally validated using system models.
5. Results from MBSE must be made available over the whole product lifecycle.
6. MBE requires models with meaning.
7. The MBE toolchain must rely on technology-independent standards.
8. Increasing complexity of products and production systems asks for new development processes, methods, and tools.
9. MBE requires changes in organization, methodology, technology, and education.
10. Investments in MBE can deliver an ROI of 3:1.

You can read the details about each of the theses in the paper. They reveal the importance of MBE for the engineering of complex products, the challenges, and the increasing merge of the two disciplines MBSE and PLM.

INCOSE defines Model-Based Systems Engineering (MBSE) as

Model-Based Systems Engineering (MBSE) is the formalized application of modeling to support system requirements, design, analysis, verification and validation activities beginning in the conceptual design phase and continuing throughout development and later life cycle phases. [INC07]

Models supersede office documents (text documents, sheets, figures) to specify products. The increasing complexity of the products and the engineering processes require the change from a document-based to a model-based approach in systems engineering.

Doing systems engineering with documents is like doing mechanical engineering with MS Paint. What you see is what you get! You have no option to process the information automatically or to create specific views for the different stakeholders.

INCOSE states in the INCOSE Vision 2020 [INC07]

In many respects, the future of systems engineering can be said to be "model-based." A key driver will be the continued evolution of complex, intelligent, global systems that exceed the ability of the humans who design them to comprehend and control all aspects of the systems they are creating.

The current INCOSE Vision 2025 [INC14] states

Model-based systems engineering will become the "norm" for systems engineering.

PLM - the other part of MBE - is also a well-established and mature discipline. Traditionally, it has a focus on mechanical engineering. However, a PLM system should manage all product data and provide services for the data in all phases of the product lifecycle (figure 4.39) [INC15].

Concept	Development	Production	Utilization / Support	Retirement

Figure 4.39: ISO 15288 generic life cycles

PLM has its origins in the world of Computer Aided Design (CAD). The holistic approach came up in the 1990ies. You must differentiate between a PLM system, i.e., a software product to provide PLM functionality, and PLM as a discipline. That is often mixed up. It is the same difference as between the discipline MBSE and MBSE modeling tools.

PLM and MBSE have separate roots, but since some years they have started to merge. The paper mentioned above is one example of this movement [PLM4MBSE15].

A crucial part of engineering is to establish an ecosystem of different cooperating methodologies and cooperating tools. No single methodology and tool cover every aspect necessary in an engineering project. On the other side, a characteristic of complex units is many relationships of different kinds. Therefore, you need a collaboration environment of methods and tools.

Open Services for Lifecycle Collaboration (OSLC) is a promising technology for a tool ecosystem on the new engineering playfield. OSLC is a set of specifications to integrate tools and their data based on standard internet technologies. The specifications, development environments, a list of tools with OSLC support, and more information about OSLC is available on the official website open-services.net.

A typical mistake in modeling is the tendency to model too many details. Over-modeling is wasted effort. The question "What should be modeled?" can be answered if you start on the other side. You create a model because some stakeholders are interested in that information. Who are the stakeholders, and which information do they need in which format?

Answer these questions, and you have the reference what you have to model. Similar to the test-driven development (TDD) approach in software engineering [Be02], you do a query-driven modeling approach. First, specify the query to retrieve the data from the model for the stakeholders, then model until the query returns the correct results.

Figure 4.40 depicts the Query-Driven Modeling (QDM) approach.

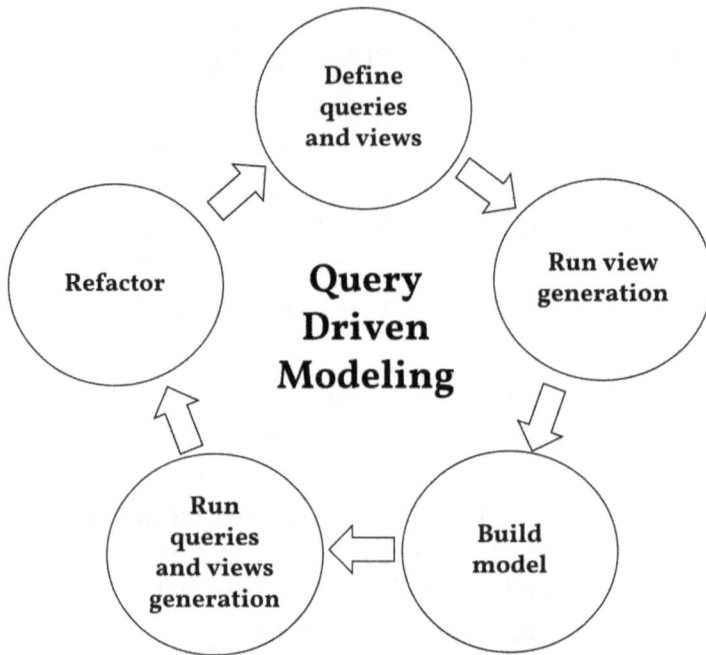

Figure 4.40: Query-Driven Modeling

First, you define the queries and views for the stakeholders, for example, a table, a document, a diagram, a traceability matrix, and so on. In the next step, you run the queries and views generators. The queries fail, i.e., create empty views, because the model data is not yet there. But it proves if the queries and views generation scripts basically work. The third step requires the real work: create the model data accordingly to the defined modeling methodology (for example, see section

4.9).

Again, run the queries and generate the views to see if it works now. Repeat the previous step until the generated views are as specified.

Before you add additional views, you should refactor the model, if necessary, i.e., adapt the model structure, move common elements into a model library, remove unused model elements, and so forth. Based on a clean model it is much easier to add new or update existing things.

While there are not many methodologies for MBE around, you find several methodologies for MBSE like Object Process Methodology (OPM) [Do16], Object-oriented Systems Engineering Method (OOSEM) [Es08], and the Systems Modeling Toolbox (SYSMOD) [We16a].

Good moves on the engineering playfield

- Store your engineering data in model repositories to enable data analysis, transformation, and the generation of stakeholder-specific views.
- Establish a collaborative engineering environment.
- Prefer standards to benefit from third-party applications, model interchange, and experiences of other organizations that use the same standard.
- Create models with the query-driven modeling approach (QDM).

4.12 REThink 4.0

Requirements engineering is a crucial part for the success of a project. We have known that for decades. Nevertheless, still many projects fail due to problems with requirements. It is time for a different way of requirements engineering: rethink and adopt a new paradigm

Requirements are often neglected in projects. They have an unagile, bureaucratic touch and the rush-to-product syndrome puts the focus on tasks that create "real" outcome, i.e., visible, working product parts and features.

On the other side, projects fail because they spend too much time with requirements. In the end, you do not sell a perfect requirement specification. Only the product counts. As always the world is not black and white.

On the new engineering playfield, a fundamental problem with requirements is the strong focus on text. That makes it nearly impossible and in any case effortful to manage hundreds or thousands of text fragments in a complex and dynamic environment where requirements often change.

An alternative is model elements representing requirements. Think, for example, of a modeled state machine - perhaps even executable - that covers exactly the desired behavior of the product. It does not make sense to translate the state machine behavior to textual requirements. The state machine itself or its parts are already requirements.

In mechanical engineering, the engineers derive a CAD model from stakeholder inputs. Those CAD models are the requirements specification for the people on the next level who really build the thing. In systems engineering, we derive, for example, a SysML model from the stakeholder inputs. Often that model is not the requirements specification for the next level. Instead, again textual requirements were derived from the model as an input for the next level. That is not very effective. Textual requirements are not as powerful as a model, and the transformation process from model to textual requirements is flawed.

Even the single transformation step from textual stakeholder needs to a model is effortful and errorful. Product engineering is an iterative process, and you have many updates of the requirements and related artifacts.

Instead of translating model elements to textual requirement statements back and forth, the model elements itself are the requirements. You can assign all requirement attributes like an ID, priority, obligation, and so on to them.

If helpful, you can also add an explanatory text statement. From such a model you still can generate pure textual requirements specification documents if you like. The document is only a view of the master data in the model. At the same time, you can have a more precise model-based definition of the requirements. Especially for functional requirements, natural languages are typically not powerful and precise enough.

That approach is called *REThink 4.0* [We15a]. The number 4 stands for the fourth stage of requirements specification.

The first stage is requirements written as simple text, i.e., textual descriptions and statements that specify the product. Such a requirements specification is easy to read - like a book -, but typically not precise and easy misunderstandable. Natural languages open space for different interpretations. Additionally, it is hard to manage the many cross-links about different aspects of the system.

The second stage is requirements written in structured text, for example, tables in a text document. The structure predefined in table rows or columns helps to cover relevant information and to manage relationships between requirements.

The third stage is textual requirements embedded in a model, for example, SysML v1 requirements or requirements in a requirements management tool like IBM DOORS. The core of the requirements is still text, typically, written in natural language. Not every aspect of a product can be precisely and well specified with text. For example, it is easier to specify a coffee cup structure with a simple CAD model than with textual statements.

The fourth stage is requirements represented by model elements like described above.

For example, Figure 4.41 depicts the definition of the execution order of the tasks A, B, C, and D in a SysML activity diagram.

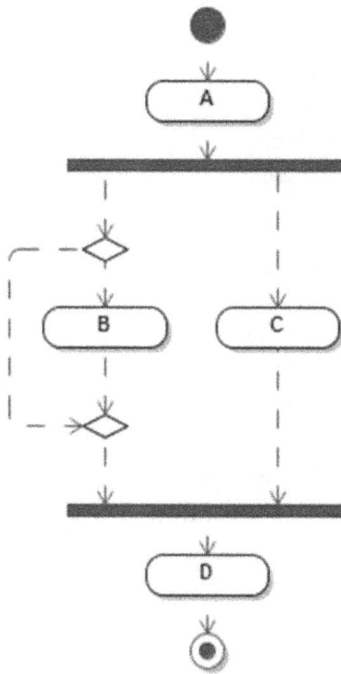

Figure 4.41: Execution order specification ABCD with SysML

It is a simple and clear specification. On the other side, it is not that clear to specify that execution order by text. The following textual specification seems to be simple and clear on the first view, too:

First A, then B and C in a random order, whereas B is optional, and finally do D.

However, if I gave people this example the other way round, i.e., first the textual specification, some people have different interpretations of the execution order. That did not happen with the model-based specification.

The recommendation is not to completely abandon the textual requirements. Modeled requirements are only used where it makes sense. If it is more effective to specify the requirement purely textually, you should do it naturally.

REThink 4.0 is a new paradigm in requirements engineering. Although the basic idea is simple, it has many consequences on the operational level and is a real change for some organizations.

Good moves on the engineering playfield

- If possible avoid requirements to communicate that information from one unit to another and instead co-locate the people.
- Avoid redundancies and use model elements as requirements instead of translating them to textual requirement statements.

4.13 Agile and Lean Systems Engineering

Agile is a common paradigm, in particular in the software development discipline. However, it is also applied outside of the software world in other engineering disciplines or any other kind of organizational unit.

There are many different opinions on what it means to be agile. I spend no effort in arguing who is right or wrong and give my understanding of agile.

First, agile is a general English word and means to be able to move quickly and easily [Ox17]. In the context of software engineering, the agile manifesto for software development defines what it means to be *agile* [Ag01].

The Agile Manifesto for Software Development was published by several well-known engineering experts (mainly software) in 2001. It is not a definition of a term or a process. It postulates 4 values and 12 principles. Organization or teams that follow those values and principles are agile.

The 4 values of the Manifesto for Agile Software Development are:

- Individuals and interactions over processes and tools
- Working software over comprehensive documentation
- Customer collaboration over contract negotiation
- Responding to change over following a plan

It is essential to understand the word "over". That word does not mean "instead". Of course, processes and tools, documentation, contract negotiations, and plans are important. The items on the right have value, but you should value the items on the left more.

The 12 principles of the manifesto give guides how to apply the values in practice:

- We follow these principles: Our highest priority is to satisfy the customer through early and continuous delivery of valuable software.
- Welcome changing requirements, even late in development. Agile processes harness change for the customer's competitive advantage.
- Deliver working software frequently, from a couple of weeks to a couple of months, with a preference to the shorter timescale.
- Business people and developers must work together daily throughout the project.
- Build projects around motivated individuals. Give them the environment and support they need, and trust them to get the job done.
- The most efficient and effective method of conveying information to and within a development team is face-to-face conversation.
- Working software is the primary measure of progress.
- Agile processes promote sustainable development. The sponsors, developers, and users should be able to maintain a constant pace indefinitely.
- Continuous attention to technical excellence and good design enhances agility.
- Simplicity–the art of maximizing the amount of work not done–is essential.
- The best architectures, requirements, and designs emerge from self-organizing teams.

- At regular intervals, the team reflects on how to become more effective, then tunes and adjusts its behavior accordingly.

The agile movement started in the software development domain, but more and more other parts of an organization adopt the agile concepts including product engineering.

However, some adoptions are necessary for that since the agile manifesto explicitly addresses software. The simplest adoption is to rename in the list of values and principles the word *software* to *solutions*, for example, the value *Working software over comprehensive documentation* to *Working solutions over comprehensive documentation.*

That works well, but something is missing. The manifesto does not address specific challenges of the product engineering domain. Together with Arie van Bennekum (co-author of the agile manifesto) and some other people I've started an initiative to write a foundation specifically for product engineering. We concluded that one part is the tweaked agile manifesto (replacing the word *software* with *solutions*). In addition, we postulate four values that address specific challenges of product engineering:

Foundation for Complex Systems Engineering

To provide a platform for continuous improvement of the development approach for complex systems...

...we value...

- Multifunctional teams over Engineering silos
- Focus on purpose over Focus on requirements
- Empowered teams over Tasked individuals

- Early learning over Late failures

(www.agile-systems-engineering.com)

Another related, but a different approach is lean systems engineering [Op11]. It is the application of Lean Thinking to the systems engineering discipline. Lean Thinking originated at Toyota means - simply said - adding value to the customer and at the same time removing waste from all activities. Taiichi Ohno, the father of Lean, said: *All we are doing is looking at the timeline from the moment the customer gives us an order to the point when we collect the cash. And we are reducing that time line by removing the non-value added wastes.* [Oh88]

The best reading about Lean is the classic book *Lean Thinking* by Womack and Jones [WJ96]. Here, I cover only the basic concepts of Lean: value and waste, and lean principles about how to create value without waste.

Value is everything that is of value for the customer, i.e., what the customer is willing to pay for or is important for him for any other reason.

Waste is anything that is not necessary to create the value, i.e., you can remove it without reducing the value.

It is not easy to identify the waste. First, naturally, you would not do anything that you think is not of value. Second, you would not like to get the response from your customers that you mistakenly categorized value as waste.

The six lean principles *Value, Map the Value Stream, Flow, Pull, Perfection,* and *Respect for People* support the process to create value without waste.

Value - Identify the value of the product from the perspective of the customer. The customer finally pays for the product. Capturing the value is a continuous process since the customer's value expectaction can change over time.

Map the Value Stream - Define your engineering process by using the customer's value as a reference point and identify all tasks that contribute to the value. All other tasks are waste. Tasks that do not add value, but are necessary should be reduced as much as possible. Tasks that do not add value and are unnecessary should be eliminated.

Flow - Perform the tasks smoothly without stopping, delays, or unplanned backflow and rework. Sometimes the unwanted backflow is interpreted as a waterfall process approach. But lean encourage iterations and methodologies with a fail early and fail often approach. That is an important learning process in particular in a cross-functional environment, and finally reduces the time to achieve the value.

Pull - A need should justify every engineering task, and the task should be completed when the internal or external customer needs the output. Creating outputs for the shelf can make the output obsolete by changing requirements, or increased effort by reactivating the output again, for example, restart discussions to understand the output.

Perfection - Improvement of the processes and products is a never-ending task. The organization should establish a learning culture and the value of perfection to get a little bit better day by day.

Respect for People - The people are the most important resource of an engineering organization. To improve processes and products it is important to learn and to make mistakes. Analyzing mistakes is about learning and not about blaming people. Chapter 3.6 discusses the resurrection of craftsmanship and the importance of people over processes and tools.

Uncertainty and continuous change are not exceptions, but daily business. It is important that the organization and the engineering processes can deal with that situation and welcome changes without starting emergency task forces. The organization must be able to respond fast and easy to changes like changing user needs, new technologies, new competitors, changed market rules, and so forth.

Agile and lean approaches enable organizations to succeed in this environment.

Good moves on the engineering playfield

- Promote a culture that lives the agile values of the (tweaked) agile manifesto and the foundation for complex systems engineering.
- Align your engineering activities with the agile and lean principles.

Bibliography

[Ag01] Agile Manifesto for Software Development. www.agil emanifesto.org.

[Al06] Dr. Nayef R.F. Al-Rodhan, Gérard Stoudmann. Definitions of Globalization: A Comprehensive Overview aand a Proposed Definition. Geneva Centre for Security Policy. 2006.

[Ar65] L. Bruce Archer. Systematic Method for Designers. Council of Industrial Design, H.M.S.O., 1965.

[As98] T.S. Ashton. The Industrial Revolution 1760-1830. Oxford University Press. 1998.

[Ba15] Ken Ball. The Dawn of the Programmable Logic Controller (PLC). www.automation.com/pulse/2015-spring. PULSE Spring Edition 2015.

[Be02] Kent Beck. Test-Driven development by Example. AWP. 2002.

[BMM15] Object Management Group. Business Motivation Model. formal/15-05-19. 2015.

[BPMN13] Object Management Group. Business Process Model And Notation. formal/13-12-09. 2013.

[Br09] Tim Brown. Change by Design: How Design Thinking Transforms Organizations and Inspires Innovation. Harper-Business. 2009.

[Br15] Jason Brownlee, Tarek Masoud, Andrew Reynolds. The Arab Spring: Pathways of Repression and Reform. Oxford

University Press. 2015.

[Co68] Conways Law. www.melconway.com/Home/Conways _Law.html. accessed June 2016.

[Co04] James O. Coplien, Neil B. Harrison. Organizational Patterns of Agile Software Development. Prentice Hall. 2004.

[Cr18] Rob Cross, Scott Taylor, Deb Zehner. Collaboration Without Burnout. Harvard Business Review. Issue July-August 2018.

[Dia11] Jared Diamond. Collapse: How Societies Choose to Fail or Succeed. Revised edition. Penguin Books. 2011.

[Do16] Dov Dori. Model-Based Systems Engineering with OPM and SysML. Springer. 2016.

[Ec10] The Economist. Too many chiefs. www.economist.com/ node/16423358. accessed November 2017.

[Ec16] The Economist. The third industrial revolution. www. economist.com/node/21553017. accessed January 2016.

[Ec17] The Economist. Britain's lonely high-flier. www.eco-nomist.com/node/12887368. accessed March 2017.

[En16] Ryan Engelman. The Second Industrial Revolution, 1870-1914. ushistoryscene.com/article/second-industrial-rev-olution/.accessed May 2016.

[Er17] Willibald Erhart. digitale Geschäftsmodelle und schnelle Innovationszyklen in der traditionellen Industrie. Master Thesis. FH Campus02. 2017.

[Es08] Jeff A. Estefan. Survey of Model-Based Systems Engineering (MBSE) Methodologies. INCOSE MBSE Initiative. 2008.

[Ga15] Oliver Gassmann, Karolin Frankenberger, Michaela

Csik. The Business Model Navigator: 55 Models That Will Revolutionise Your Business. FT Press. 2015.

[Ga18] Oliver Gassmann, Karolin Frankenberger, Michaela Csik. The St. Gallen Business Model Navigator. Working Paper. University of St.Gallen. www.bmi-lab.ch. accesses March 2018.

[Gar18] Gartner. Glossary. Bimodal. accessed November 2018. https://www.gartner.com/it-glossary/bimodal/.

[Ger14] Federal Ministry of Education and Research (BMBF), The new High-Tech Strategy Innovations for Germany, 2014.

[Gu12] The Guardian. The rise of the meaningless job title. www.theguardian.com/money/work-blog/2012/mar/02/rise-of-meaningless-job-title. accessed November 2017.

[He16] Heise. Der Vater des Autokonsums. www.heise.de/autos/artikel/Der-Vater-des-Autokonsums-1926988.html. accessed Januar 2016.

[INC07] INCOSE Technical Operations. 2007. Systems Engineering Vision 2020, version 2.03. Seattle, WA: International Council on Systems Engineering, Seattle, WA, INCOSE-TP-2004-004-02.

[INC14] International Council on Systems Engineering (INCOSE). Systems Engineering Vision 2025. 2014. www.incose.org/docs/default-source/aboutse/se-vision-2025.pdf?sfvrsn=4. accessed June 2018.

[INC15] International Council on Systems Engineering (INCOSE). Systems Engineering Handbook. 4th edition. 2015.

[INC16] International Council on Systems Engineering. What is Systems Engineering?. www.incose.org/AboutSE/WhatIsSE.

accessed March 2016.

[Ind17] The Independent. Missing flight MH370. www.independent.co.uk/news/world/asia/missing-flight-mh370-rolls-royce-dragged-into-the-mystery-as-rumours-surface-suggesting-that-data-9190622.html. accessed March 2017.

[I413] Communication Promoters Group of the Industry-Science Research Alliance and acatech. Securing the future of German manufacturing industry: Recommendations for implementing the strategic initiative INDUSTRIE 4.0 - Final report of the Industrie 4.0 Working Group. 2013.

[ISO15288] ISO/IEC/IEEE 15288. (2015). Systems and Software Engineering—System Life Cycle Processes. Geneva, Switzerland: International Organization for Standardization.

[Ko16] Pawel Korzynski, Elizabeth Florent-Treacy, Manfred F.R. Kets de Vries. You and Your Technostress: Relating Personality Dimensions to ICT-Related Stress. INSEAD Working Paper No. 2016/31/EFE. 2016.

[LAT16] Los Angeles Times. German automakers who once laughed off Elon Musk are now starting to worry. www.latimes.com/business/autos/la-fi-hy-0419-tesla-germany-20160419-story.html. April 19th, 2016. accessed June 2016.

[Le96] Harry Levinson. When Executives Burn Out. Harvard Business Review. July-August 1996.

[Le08] Edward A. Lee. Cyber Physical Systems: Design Challenges. Technical Report No. UCB/EECS-2008-8. www.eecs.berkeley.edu/Pubs/TechRpts/2008/EECS-2008-8.html.

[Luh87] Niklas Luhmann. Soziale Systeme - Grundriß einer allgemeinen Theorie. Suhrkamp. 1987.

[Mea04] Donella H. Meadows, Jorgen Randers, Dennis L.

Meadows.Limits to Growth: The 30-Year Update. 3rd edition. Chelsea Green Publishing. 2004.

[Mo89] David Montgomery. The Fall of the House of Labor: The Workplace, the State, and American Labor Activism, 1865-1925. Cambridge University Press. 1989.

[Na16] The National Archives, www.nationalarchives.gov.uk/ education/politics/g5/, accessed January 2016.

[OCEB11] Tim Weilkiens et al. OCEB Certification Guide. Morgan Kaufmann. 2011.

[Oh88] Taiichi Ohno. The Toyota Production System: Beyond Large Scales Production. Productivity Press. 1988.

[Op11] Bohdan W. Oppenheim. Lean for Systems Engineering with Lean Enablers for Systems Engineering. Wiley. 2011.

[Os10] Alexander Osterwalder, Yves Pigneur. Business Model Generation: A Handbook for Visionaries, Game Changers, and Challengers. Wiley. 2010.

[Os14] Alexander Osterwalder, Yves Pigneur, Gregory Bernarda, Alan Smith, Trish Papadakos. Value Proposition Design: How to Create Products and Services Customers Want. Wiley. 2014.

[Ox17] Oxford Dictionary Online, www.oxforddictionaries.com, accessed 2017.

[Pa87] Patent US372786: Gramophone. Published 1887.

[Pe96] Chris Peers. Imperial Chinese Armies (2), 590-1260 AD. Osprey Publishing. 1996.

[Pi16] Roman Pichler. Strategize: Product Strategy and Product Roadmap Practices for the Digital Age. Pichler Consulting. 2016.

[PLM4MBSE15] PLM4MBSE Working Group Position Paper.

10 theses about MBSE and PLM - Challenges and Benefits of Model Based Engineering (MBE). GfSE - German Chapter of INCOSE. 2015.

[Re14] Die Welt, "Wir verdienen auch mit unseren Elektroautos Geld", www.welt.de/wirtschaft/article125475868/Wir-verdienen-auch-mit-unseren-Elektroautos-Geld.html, accessed January 2018.

[Rif13] Jeremy Rifkin. The Third Industrial Revolution: How Lateral Power Is Transforming Energy, the Economy, and the World. St. Martin's Griffin. 2013.

[Ro69] Bernd Rohrbach. Kreativ nach Regeln – Methode 635, eine neue Technik zum Lösen von Problemen. Absatzwirtschaft. 12: 73–75. 1969.

[SBVR17] Object Management Group. Semantics Of Business Vocabulary And Rules. formal/17-05-05. 2017.

[Si13] Kaushik Sinha, Olivier de Weck. Structural Complexity Quantification for Engineered Complex Systems and Implications on System Architecture and Design. Proceedings of the ASME Design Engineering Technical Conference. 3. 10.1115/DETC2013-12013.

[SMLC17] Smart Manufacturing Leadership Coalition. What is Smart Manufacturing?. www.cmtc.com/blog/what-is-smart-manufacturing-part-1a-of-6. accessed March 2017.

[Sn07] David J. Snowden, Mary E. Boone. A Leader's Framework for Decision Making. Harvard Business Review. November 2007.

[Sn16] SnapCash. support.snapchat.com/en-US/ca/snapcash. accessed June 2016.

[SysML17] Object Management Group: OMG Systems Modeling Language (OMG SysML), Version 1.5. formal/2017-05-01.

[SW18] Olivier de Weck. The First Law of Systems: Conservation of Complexity. SWISSED Conference 2018.

[Tay11] Frederick Winslow Taylor. The Principles of Scientific Management. New York, Norton. 1947.

[UML17] Object Management Group: Unified Modeling Language (UML), Version 2.5.1. formal/2017-12-05.

[We15] Tim Weilkiens, Jesko G. Lamm, Stephan Roth, Markus Walker. Model-Based System Architecture. Wiley. 2015.

[We15a] Tim Weilkiens. REThink 4.0 - Requirements Engineering with MBSE. SWISSED conference 2015.

[We16a] Tim Weilkiens. SYSMOD - The Systems Modeling Toolbox. 2nd edition. MBSE4U. 2016.

[We16b] Tim Weilkiens, Christian Weiss, Andrea Grass, Kim Duggen. OCEB 2 Certification Guide: Business Process Management - Fundamental Level. 2nd edition. Morgan Kaufmann. 2016.

[We18] Tim Weilkiens. SYSMOD Workshop Cards. www.mbse-4u.com/sysmod-workshop-cards

[WJ96] J. P. Womack, D. T. Jones. Lean Thinking. Simon & Shuster. 1996

[Woh04] Gerhard Wohland, Judith Huther-Fries, Matthias Wiemeyer, Dr. Jörg Wilmes. Vom Wissen zum Können - Merkmale dynamikrobuster Höchstleistung. DETECON. 2004

Image references

[ImgBe] Emil Berliner. Photograph taken 1887. Public domain. commons.wikimedia.org/wiki/File:Gramophone_berliner2.jpg

[ImgFA] Ford assembly line. 1913 photograph Ford company, USA. Public domain. commons.wikimedia.org/wiki/File:Aline1913.jpg

[ImgFT] Frederick Winslow Taylor. Public domain. commons.wikimedia.org/wiki/File:Frederick_Winslow_Taylor.JPG

[ImgHF] Henry Ford. Time Magazine 1935. Photo credit: Jeffrey White Studios, Inc. Public Domain. commons.wikimedia.org/wiki/File:Timehenryford-crop.jpg

[ImgPL] Power Loom. Textile Mercury newspaper 1892 Issue 158. Public Domain. commons.wikimedia.org/wiki/File:TM158_-Strong_Calico_Loom_with_Planed_Framing_and_Catlow%27s_-Patent_Dobby.png

[ImgSB] Jean-Pierre Houël (1735–1813). Bibliothèque nationale de France. Public Domain. commons.wikimedia.org/wiki/File:Prise_de_la_Bastille.jpg

[ImgTM] Threshing Machine. Drawing of a horse-powered thresher from a French dictionary (published in 1881). Public Domain. commons.wikimedia.org/wiki/File:Batteuse_1881.jpg

Index

www.ingramcontent.com/pod-product-compliance
Lightning Source LLC
Chambersburg PA
CBHW021559210326
41599CB00010B/517